Planetary Science - Explorations, Missions, and Mapping
http://dx.doi.org/10.5772/intechopen.1009033
Edited by Yann-Henri Chemin

Contributors
Hector Pérez-de-Tejada, Marina Shcherbina, Nikolai Kiselev, Yann-Henri Chemin, Yann H. Chemin, Şengül Yalgın

First published in London, United Kingdom, 2026 by IntechOpen
IntechOpen is the global imprint of INTECHOPEN LIMITED, registered in England and Wales, registration number: 11086078, 167-169 Great Portland Street, London, W1W 5PF, United Kingdom

For EU product safety concerns: IN TECH d.o.o., Prolaz Marije Krucifikse Kozulić 3, 51000 Rijeka, Croatia, info@intechopen.com or visit our website at intechopen.com.

British Library Cataloguing-in-Publication Data
A catalogue record for this book is available from the British Library

Planetary Science - Explorations, Missions, and Mapping
Edited by Yann-Henri Chemin
p. cm.
Print ISBN 978-0-85466-950-9
Online ISBN 978-0-85466-949-3
eBook (PDF) ISBN 978-1-83769-999-5

Meet the editor

Dr. Yann-Henri Chemin has a Ph.D. in Algorithmic Remote Sensing Applications from the Asian Institute of Technology, TH, 2006, where he worked on the thermodynamic energy balance on the Earth's surface. He then graduated in planetary sciences with an astronomy specialization in 2016 from Birkbeck, University College London, UK, where he worked on the hyperspectral analysis of the Apollo 12 landing site and its vicinity on the Moon. He also worked on the surface–atmosphere interactions of the Moon Titan, and on the mapping of craters on the dwarf planet called Ceres. He then completed an EMBA in Security, Defense, and Space in 2021 at the University of Nice-Côte d'Azur (France). Dr. Chemin's interests include challenges of settling on the Moon and Mars, especially food creation, settlement construction, and in situ resource utilization. He edited two books on lunar science, a chapter on the state of the New Space/Moon race, one book on the ionosphere, and one on the solar system and cryovolcanism.

Contents

Preface

The exploration of planetary bodies beyond Earth has transformed our understanding of the solar system's evolution, composition, and dynamic processes. This edited volume brings together four complementary studies that examine diverse aspects of planetary science, from the hydrocarbon cycles shaping Titan's unique environment to the dust enrichment pathways connecting stellar death with planetary formation, and from the plasma physics governing Venus's ionosphere to the polarimetric signatures revealing surface properties of Jupiter's and Saturn's moons.

The introductory chapter reviews the science covered in this contribution. It looks at how this is meaningful in the context of planetary science and its necessary vectors, the instruments, missions, and spacecraft, as they are deployed in recent years, and yet to arrive at their operational destination.

The second chapter, "A Review of Hydrocarbon Cycles and Their Impacts on Titan's Surface", provides a comprehensive synthesis of our current knowledge of Titan's methane-based meteorology and surface processes. Saturn's largest moon stands alone in the solar system with its dense nitrogen-rich atmosphere and active hydrocarbon cycle analogous to Earth's water cycle. Drawing upon findings from the Cassini mission, ground-based observatories, and recent James Webb Space Telescope observations, this review examines how methane and ethane cycles shape Titan's landscape through precipitation, evaporation, fluvial erosion, and sediment transport. The chapter explores the energy balance driving these processes, the formation of vast methane-ethane lakes and seas, and the complex organic chemistry occurring in Titan's atmosphere. Spectroscopic detections of diverse molecules, including ethyl cyanide, vinyl cyanide, and cyclopropenylidene, underscore Titan's potential as a natural laboratory for prebiotic chemistry.

The third chapter, "Variability of Polarization on the Surface of Jupiter's and Saturn's Moons", presents high-precision polarimetric observations revealing longitudinal variations in surface properties across multiple satellites. Linear polarization measurements obtained with the POLSHAKH polarimeters at the Crimean Astrophysical Observatory demonstrate that Europa, Io, Ganymede, Callisto, Iapetus, and Rhea exhibit longitude-dependent polarization superimposed on their phase-angle dependence. These variations, detected at levels of a few hundredths of a percent, indicate optical and structural inhomogeneities resulting from volcanic activity, magnetospheric processing, meteoritic bombardment, and compositional differences between leading and trailing hemispheres. For Io, temporal changes in the longitudinal polarization pattern may reflect variations in global volcanic activity over decadal timescales.

The fourth chapter, "Planetary Nebulae as Agents of Interstellar Enrichment and Stellar Evolution Insight", bridges stellar evolution and planetary system formation by examining how dying stars return processed material to the interstellar medium.

Planetary nebulae represent a crucial but brief phase in the life cycle of intermediate-mass stars, during which chemically enriched gas and dust are ejected into space. Using photoionization modeling of sixteen well-studied planetary nebulae, this chapter quantifies dust-to-gas mass ratios, total ejected masses, and grain abundance relative to the interstellar medium. The analysis reveals diverse dust compositions, from carbon-rich to silicate-dominated, reflecting differences in progenitor chemistry and subsequent processing. The chapter traces how this stellar ejecta is incorporated into molecular clouds and, eventually, into protoplanetary disks, establishing a direct link between the final stages of stellar evolution and the raw materials available for planet formation.

The fifth chapter, "Mean Free Path Values of Wave-Particle Interactions in the Venus Ionosheath and in the Venus Upper Ionosphere", applies fluid dynamics principles to understand momentum transfer from the solar wind to Venus's upper atmosphere. Unlike Earth, which has a protective magnetic field, Venus's ionosphere directly interacts with the solar wind, producing a complex boundary region. This study derives mean free path values for wave-particle interactions in both the ionosheath and the upper ionosphere using relationships among thermal speed, viscosity, and dissipation. The analysis reveals that mean free path values decrease from the ionosheath toward the upper ionosphere, where wave-particle interactions gradually transition to Coulomb collisions at lower altitudes. These findings provide quantitative constraints on the trans-terminator flow that redistributes plasma from Venus's dayside to nightside.

Together, these chapters demonstrate the power of combining diverse observational techniques with theoretical modeling to unravel planetary processes. From Titan's hydrocarbon lakes to the dust grains expelled by dying stars, from Venus's ionospheric dynamics to the polarimetric signatures of icy satellites, each study contributes essential pieces to our understanding of how planetary bodies form, evolve, and interact with their environments. The interdisciplinary nature of these investigations reflects the modern approach to planetary science, where insights from atmospheric physics, stellar evolution, plasma dynamics, and remote sensing converge to create a more complete picture of our solar system.

We hope this volume will serve as a valuable resource for researchers, graduate students, and anyone interested in the ongoing exploration and characterization of planetary bodies across our solar system.

Yann-Henri Chemin
Seilio Douar E.I.,
Plumergat, France

Section 1

Introduction

Chapter 1

Introductory Chapter: Planetary Science – Explorations, Missions and Mapping

Yann-Henri Chemin

1. Explorations and key missions

Planetary science encompasses the study of planets, moons, asteroids, comets, and other solar system bodies through a combination of robotic exploration, remote sensing, *in situ* measurements, and advanced numerical modeling. This discipline has rapidly advanced over recent decades, driven by increasingly capable spacecraft missions and ground-based observatories that reveal the formation, evolution, and dynamic processes shaping these worlds. Key themes include the diversity of planetary environments, from rocky inner planets to gas and ice giants, and the search for habitable conditions across ocean worlds and beyond.

Planetary exploration began with flyby missions in the 1960s and 1970s, such as Mariner and Voyager, which provided the first close-up views of multiple planets and established the foundational framework for comparative planetology. These early efforts evolved into extended orbital surveys, landers, and rovers that delivered high-resolution imaging, spectroscopy, and geochemical analyses. Long-duration missions like Mars Reconnaissance Orbiter (in operation since 2006), Lunar Reconnaissance Orbiter (2009–present), and New Horizons (launched in 2006) continue to yield transformative data on surface geology, atmospheric dynamics, and volatile inventories.

Past missions to dwarf planets have profoundly expanded our understanding of the Kuiper Belt and primitive Solar System bodies. NASA's New Horizons conducted the first reconnaissance of Pluto in July 2015, revealing a geologically active world with nitrogen ice plains, water ice mountains, and organic hazes in a tenuous atmosphere, fundamentally reshaping models of dwarf planet diversity [1]. The mission's subsequent flyby of Arrokoth (486958) in 2019 sampled a pristine contact binary Kuiper Belt object, providing direct evidence of gentle pebble-cloud accretion during planet formation.

A primary focus of contemporary exploration targets ocean worlds, where subsurface liquid water may harbor conditions suitable for life. NASA's Europa Clipper mission, launched in 2024, will conduct dozens of close flybys of Jupiter's moon Europa to map its icy surface, probe the underlying ocean with ice-penetrating radar, and assess geochemical exchanges via spectroscopy and mass spectrometry [2]. Complementarily, the European Space Agency's JUICE (JUpiter ICy moons Explorer), arriving in 2031, will orbit Ganymede – the largest moon in the Solar

System – for detailed studies of its unique magnetic field, internal structure, and geological history, while also surveying Callisto and Europa [3].

Upcoming icy moon missions further intensify focus on Saturn's ocean worlds. NASA's proposed Enceladus Orbilander would combine repeated orbital remote sensing of plume chemistry with targeted surface landings to sample organics and salts from the subsurface ocean [4]. The Enceladus Life Finder (ELF) concept, building on Cassini discoveries of molecular hydrogen and silica nanoparticles in plumes, would aim to detect biosignatures through high-sensitivity mass spectrometry. NASA's Dragonfly rotorcraft-lander (2034 arrival at Titan) pioneers aerial mobility to sample diverse organic dunes, lakes, and mountains, building on Cassini–Huygens discoveries of a methane-driven hydrological cycle [5].

Ice giant systems represent another frontier, with the proposed Uranus Orbiter and Probe mission prioritizing *in situ* atmospheric sampling, satellite flybys (including ocean-bearing moons Miranda, Ariel, Umbriel, and Titania), and ring characterization to address gaps in understanding ice giant formation and evolution. Mars remains a cornerstone of exploration, with the Perseverance rover (2021 landing) caching samples for eventual return to Earth, enabling isotopic and mineralogic analyses unattainable *in situ* [6].

Small-body exploration and planetary defense are also advancing [7], with missions like NASA's Psyche (launched in 2023, arrival in 2029) targeting a metal-rich asteroid to probe planetary cores, and the NEO Surveyor infrared telescope surveying near-Earth objects for hazard mitigation while cataloging primitive materials [8]. Sample return missions like OSIRIS-REx (Bennu samples returned in 2023) and Japan's Hayabusa2 (Ryugu, 2020) provide pristine materials for laboratory scrutiny. These efforts span the Solar System, from Mercury's orbiters to Kuiper Belt outposts [9], revealing a dynamic tapestry of geological activity, volatile cycling, and potential biosignatures.

2. Planetary mapping and integrated datasets

Planetary mapping synthesizes diverse observations into geospatial frameworks that reveal topography, composition, tectonics, and time-variable phenomena. Traditional photogrammetry and photoclinometry have evolved into integrated products using stereo imaging, laser altimetry (e.g., Mars Orbiter Laser Altimeter), and radar (e.g., Magellan for Venus), producing digital elevation models with resolutions down to meters. These maps delineate craters, volcanoes, faults, and depositional basins, guiding interpretations of impact flux, endogenous resurfacing, and erosion.

Compositional mapping employs multispectral and hyperspectral imagers to identify minerals, ices, and volatiles. New Horizons mapped Pluto's diverse terrains – Sputnik Planitia's convective nitrogen ice, rugged water-ice peaks, and tholin-coated highlands – revealing ongoing geological activity on a distant world. On the Moon, Lunar Reconnaissance Orbiter's Diviner instrument maps thermal inertia and rock abundance, while Mars' CRISM spectrometer reveals hydrated clays and sulfates linked to past water activity.

Atmospheric mapping tracks cloud motions, temperature profiles, and trace gases via limb sounding and thermal imaging. Titan's methane clouds and lakes have been charted through Cassini Visual and Infrared Mapping Spectrometer data, assimilated into general circulation models, revealing seasonal shifts in humidity and precipitation [10]. Venus's dynamic cloud decks are mapped using Akatsuki's IR camera, showing equatorial Kelvin waves and polar vortices that drive super-rotation.

Geophysical mapping leverages gravity and magnetic data to infer interior structures. Gravity anomalies from spacecraft tracking reveal mass concentrations (mascons) on the Moon and subsurface oceans on Ganymede, Europa, and Enceladus. Magnetic field mapping at unmagnetized bodies like Venus delineates induced magnetotails and ion escape channels, informing atmospheric evolution models.

Modern mapping is increasingly dynamic and data-assimilative, incorporating time-series observations into four-dimensional (4D) models. Machine learning enhances feature detection in vast image archives from New Horizons, Cassini, and Galileo, while standardized projections (e.g., USGS PDS cartography) facilitate cross-mission comparisons. For exoplanets, phase-curve mapping with JWST infers atmospheric circulation and heat transport. Mapping underpins mission planning: high-resolution topography selects safe landing sites, hazard maps guide rover paths, and resource maps target *in situ* utilization. As datasets proliferate, open-access portals like NASA's PDS and ESA's PSA democratize analysis, fostering global collaboration.

3. Conclusion

The rapid evolution of planetary science – driven by ambitious missions, advanced mapping technologies, and interdisciplinary collaboration – has transformed our understanding of the Solar System's diversity, from the geologically active surfaces of dwarf planets to the hidden oceans of icy moons. As upcoming missions like Europa Clipper [2], JUICE [3], and Dragonfly [10] push the boundaries of exploration, the integration of remote sensing, *in situ* analysis, and data-driven modeling promises to unlock new insights into planetary formation, habitability, and the potential for life beyond Earth. This dynamic field not only expands scientific knowledge but also inspires global cooperation and innovation, ensuring that the quest to map and understand our cosmic neighborhood continues to thrive in the decades ahead.

Author details

Yann-Henri Chemin
Seilio Douar, Plumergat, France

*Address all correspondence to: dr.yann.chemin@gmail.com

References

[1] Stern SA, et al. The Pluto system: Initial results from its exploration by New Horizons. Science. 2015;**350**: aad8866

[2] Pappalardo RT, et al. The Europa Clipper mission: Exploring Jupiter's ocean world. Space Science Reviews. 2024;**220**:6

[3] Grasset O, et al Jupiter ICy moons Explorer (JUICE): An ESA mission to orbit Ganymede and to characterise the Jupiter system. Planetary and Space Science. 2013;**78**:1–21

[4] MacKenzie SM, et al. The enceladus orbilander mission concept: Balancing return and resources in the search for life. The Planetary Science Journal. 2021;**2**:77

[5] Lorenz RD, Turtle EP, et al. (2017). Dragonfly: A rotorcraft lander concept for scientific exploration at Titan (Tech. Rep.). Johns Hopkins APL Technical Digest

[6] Farley KA, et al Mars 2020 mission overview. Space Science Reviews. 2020;**216**:142

[7] National Academies of Sciences, Engineering, and Medicine. Origins, Worlds, and Life: A Decadal Strategy for Planetary Science and Astrobiology 2023–2032. Washington, DC: National Academies Press; 2022

[8] Mainzer A, et al. The near-earth object surveyor mission. Planetary Science Journal. 2023;**4**:224

[9] Stern SA, et al. Initial results from the New Horizons exploration of 2014 MU 69, a small Kuiper Belt object. Science 2019;**364**:eaaw9771

[10] Lora JM. Moisture transport and the methane cycle of Titan's lower atmosphere. Icarus. 2024;**422**:116241

Section 2

Rocky Planets and Moons Surfaces

Chapter 2

A Review of Hydrocarbon Cycles and Their Impacts on Titan's Surface

Yann-Henri Chemin

Abstract

Titan (Saturn VI), Saturn's largest moon, is a unique planetary body in the solar system, distinguished by its dense, nitrogen-rich atmosphere and an active hydrocarbon cycle that is unparalleled elsewhere. This chapter provides information on our understanding of the hydrocarbon processes on Titan. By synthesizing findings from major missions and studies, it explores how these cycles shape Titan's surface and atmosphere. Recent advances from the Cassini mission, ground-based observatories, and the James Webb Space Telescope (JWST) have significantly expanded our knowledge of Titan's methane cycle, atmospheric chemistry, and the intricate interactions between surface and atmosphere. High-resolution observations have revealed dynamic cloud formation, seasonal precipitation events, and extensive surface changes, including evidence for fluvial erosion, sediment transport, and the presence of vast methane-ethane lakes and seas. Spectroscopic studies have identified a diverse suite of organic molecules (such as ethyl cyanide, vinyl cyanide, cyclopropenylidene, and propadiene), underscoring the complexity of Titan's organic chemistry and its potential as a natural laboratory for prebiotic processes. New theoretical models and impact studies have provided important insights into the sources and sinks of atmospheric methane, highlighting the delicate balance required to sustain Titan's methane inventory over geologic timescales. The persistence of methane and ethane, despite ongoing photolytic destruction, continues to challenge our understanding of planetary climate evolution and the processes that may have influenced early Earth-like environments.

Keywords: Titan, methane, ethane, cycle, erosion, energy balance

1. Introduction

Titan (Saturn VI) is one of the most studied icy satellites in the solar system [1]. It is the largest of Saturn's natural satellites, a rocky moon with a cloudy atmosphere, being 1.5 times the Moon's diameter and 80% more massive, its annual cycle is equivalent to 29.45 Earth years [2, 3]. It has a high content of surface and atmospheric hydrocarbon hovering around their triple phase temperature range (solid, liquid, gas).

IntechOpen

The atmosphere is composed of 94% by mass of N, alongside a significant amount of CH (5% by mass at the troposphere) [1]. The hydrocarbon and nitrile species that condense into optically thick global haze layers, is photochemically generated from the hydrocarbon methane (CH_4) of solar system origins, the works of [1, 4–6], found ethane (C_2H_6) rainfall, confirming early work by Lunine et al. [7] from the Voyager 1 radio occultation experiment.

Work on the northern summer season by Nixon et al. [3] has shown that the MIRI sensor on the JWST detected the radical methyl, which is the result of the break-up of methane, forming heavier molecules but also creating ethane. More recent work found ethyl cyanide (CH_3CH_2CN; in 2015) [8], vinyl cyanide (CH_2CHCN; in 2017) [?], and cyclopropenylidene (c–C_3H_2; in 2020) [9], or with the use of high-resolution infrared spectroscopy, as was the case with the discovery of propadiene (CH_2CCH_2; in 2019) from observations with the IRTF-TEXES [10]. Finally, recent work by Rianço-Silva et al. [11] found C_3 at the upper atmosphere of Titan, with a column density of 10^{13} cm^{-2}. This study is employing the (m)ethanological cycle, while the study by Burr et al. [12] considers the dynamics and cycling of hydrocarbons on Titan to be "Titan's alkanology."

The Earth's hydrological cycle is a conceptual analog to orientate the description of Titan alkanology. On planet Earth, a hydrological cycle description can be summarized as water transiting on land (sub)surface through gravity, returning by either sublimation or evaporation (transpiration of biological entities could be added here), accumulating into the atmosphere, and descending onto the land surface as ice/snow/rainfall. On oceans (or lakes), it can be simplified further as sublimation/evaporation complemented by the same ice/snow/rainfall as on land. Lateral exchanges do exist, like rivers and deltas connecting to oceans, main oceanic currents, and long-term transfers in aquifers. Even longer water transfers exist through plate tectonics and geological formations like andesitic ones. Geological water transfers are not considered in hydrological sciences, which are more concerned with the water needs of actual civilizations. These mostly surface changes of location and phases of water are generated by the energy transfers between the top of the atmosphere and the planetary surface. The global energy budget of a planet is thus the necessary information to understand conduction, convection, and vaporization of a condensable gas in a surface–atmosphere interaction setting.

2. Energy balance

The global energy budget of Titan is the fundamental driver behind the phase transitions of hydrocarbons in its atmosphere and on its surface. As discussed by Hayes [2], the observed rate of hydrocarbon photolysis in Titan's upper atmosphere necessitates either a significant internal source of methane and related species from the surface or subsurface, or else Titan's atmospheric methane would have been depleted over geologically short timescales. Furthermore, Hayes [2] highlights the importance of Titan's seasonal periodicity, with a Titan year spanning 29.5 Earth years, as well as the influence of long-term Milankovitch cycles on the order of 50,000 and 270,000 years, which may modulate climatic and photochemical processes.

The primary source of radiative energy for Titan, as for all planetary bodies, is its parent star. While Earth receives approximately 1,800 W m^{-2} of solar irradiance at the top of its atmosphere, Titan's exo-atmospheric irradiance is estimated to be just 15.1 W m^{-2} – a mere 1.1% of the value typical for Earth's tropics, as shown in **Table 1** [13]. This comparison likely references incoming radiation at Earth's surface for context. The transmission of this irradiance through Titan's dense, haze-laden atmosphere is subject to complex interactions, including scattering, diffraction, and reflection by atmospheric particles. These processes, together with the interplay of greenhouse and antigreenhouse effects, ultimately determine the energy budget available at the surface [14].

Upon reaching Titan's surface, a portion of the incoming shortwave (SW) solar radiation is reflected back into the atmosphere due to surface albedo effects (see **Figure 1**). Surface albedo, defined as the integrated reflectance between 0.3 and 3 μm, varies widely depending on surface composition: about 5% for liquid water oceans, 15%–25% for terrestrial vegetation, 35%–40% for sand or desert, and 60%–80% for snow or clouds on Earth. On Titan, albedo changes have been observed at low latitudes [15], with values ranging from very low (dark) regions to bright dune fields. Notably, such variations are often attributed to the presence of liquid methane on the surface, especially following runoff from intense rainstorm events. The fraction of energy reflected or absorbed at the surface is critical, as it is subsequently reprocessed by the atmosphere through additional scattering and absorption before escaping to space.

In addition to the SW component, the surface also receives longwave (LW; **Figure 1**) radiation, which is modulated by atmospheric absorption windows and the emissivity properties of surface materials. The transition from blackbody to greybody emission is governed by the surface emissivity (ε), a key parameter influencing the thermal infrared signature detected by remote sensing instruments [16]. On Earth, emissivity typically exceeds 0.9 (often 0.96–0.98), and the Stefan–Boltzmann law describes the relationship between emitted energy and temperature:

$$GE = \varepsilon \sigma T^4 \tag{1}$$

where GE is the greybody energy flux (W m^{-2}), σ is the Stefan–Boltzmann constant, and T is the surface temperature. The combined effects of SW and LW energy exchanges yield the net radiation (R_{net}) at the surface, which represents the total energy available to drive thermodynamic and phase-change processes:

$$R_{net} = SW_{bal} + LW_{bal} \tag{2}$$

Quantity	Titan global	$\left[\frac{TitanGlobal}{EarthTropics}\right]$
Number density, $n_a(\text{cm}^{-3})$	1.1×10^{19}	4.7
Heat of vaporization, $L(\text{ Jmole}^{-1})$	8.6×10^3	0.20
Solar constant, $S_0(\text{Wm}^{-2})$	15.1	0.011
Evaporative flux, $H_L(\text{Wm}^{-2})$	$\lesssim 1.27$	$\lesssim 8 \times 10^{-3}$
Bulk transfer coefficient, $C(\text{ms}^{-1})$	$\gtrsim \% 3 \times 10^{-4}$	$\gtrsim \% 0.02$
Departure from saturation, $\frac{(q_s - q_a)}{q_s}$	$\lesssim 0.02$	$\lesssim 0.1$
Lifting condensation level, (m)	$\lesssim 200$	$\lesssim 0.4$

Table 1.
Atmospheric boundary layer of Titan (methane, 95 K) after the study by Flasar [13].

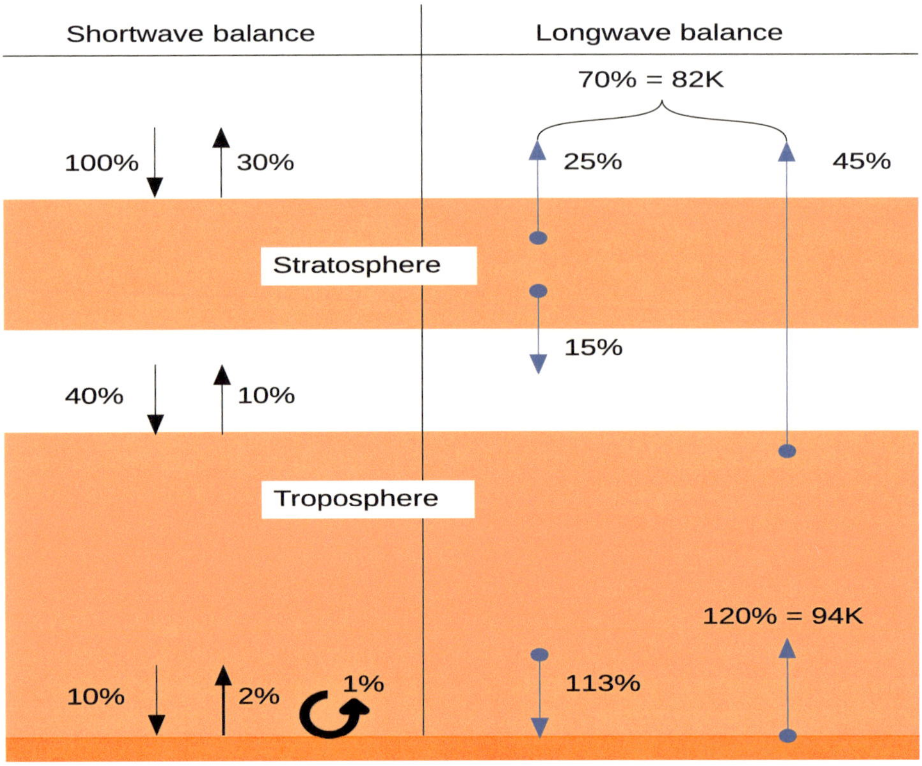

Figure 1.
Energy balance at Titan's surface-[tropo/strato]spheres, expanded from the study by McKay et al. [14]

Here, R_{net} is the net radiation budget (W m^{-2}) at the surface, and SW_{bal} and LW_{bal} are the respective SW and LW balances. On a planet such as Earth, where water exists in all three phases (gas, liquid, solid), this available energy is partitioned among phase transitions (e.g., sublimation, melting, evaporation), as well as conductive and convective heat transfers within the surface and atmosphere.

On Titan, methane (CH_4) plays a role analogous to water on Earth, acting as a condensable greenhouse gas. As shown by McKay et al. [14], the primary contributors to Titan's greenhouse effect are collision-induced absorptions involving $N_2 - N_2$, $CH_4 - N_2$, and $H_2 - N_2$ pairs. The top-of-atmosphere albedo is approximately 30%, and of the remaining 70% of incident solar energy, only about 10% actually reaches the surface. This result in a single-pass atmospheric transmissivity of roughly 14%, which is significantly lower than the typical values observed on Earth. Titan's energy budget is further complicated by the presence of both greenhouse and antigreenhouse effects, which arise from the interplay of atmospheric windows and the altitude-dependent appearance of haze and cloud layers. The energy transfer between the troposphere and stratosphere (see **Figure 1**) can be described by the following relationship [14]:

$$\varepsilon\sigma T^4 = F^- - F^+ + IR_s \tag{3}$$

In this equation, ε and T represent the combined emissivity and temperature of the surface and troposphere, F^- is the stratospheric downwelling solar flux (40%),

F^{+} is the tropospheric upwelling solar flux (10%), and IR_s is the stratospheric thermal emission directed downward (15%). All fluxes are in units of W m^{-2}, and the percentages are illustrated in **Figure 1**. This framework helps explain the observed atmospheric temperature profile, which transitions from 94 K at the surface to 71 K at 40 km altitude and rises to 170 K in the stratosphere [7, 14], and more recently as in the study by Waite et al. [17]. These complex radiative and thermal interactions underpin the unique climate and surface processes observed on Titan, distinguishing it from other bodies in the solar system.

3. Evaporation

The potential existence of hydrocarbon oceans on Titan was first hypothesized by Flasar [13] and Lunine et al. [7], following their analyses of data from the Voyager 1 radio occultation experiment. In particular, Flasar [13] based his conclusions on comprehensive assessments of Titan's radiative and energy balances, emphasizing a predicted discontinuity of approximately 1–3 K in the thermal profile at the interface between the surface and the atmosphere. This discontinuity, coupled with radio occultation data indicating that 13%–33% of incoming solar radiation is absorbed by the surface, provided a robust foundation for proposing the existence of latent heat fluxes sufficient to drive vertical upwelling. Building upon these insights, Flasar [13] further developed a latent heat flux equation relevant for an open liquid body on Titan, expressed as:

$$\mathrm{H}_L = \frac{n_a L}{\eta} C(q_s - q_a) = \frac{n_a L}{\eta} C\delta q \tag{4}$$

where η denotes Avogadro's number, L is the latent heat of vaporization of methane, C is the bulk transfer coefficient, and q represents the mole fraction of methane at the liquid surface (saturated state, q_s) and at a reference height in the atmosphere above (q_a). The triple point of methane (CH_4) occurs at 91 K and 0.2 bar, while Titan's surface conditions are typically around 94 K and 1.47 bar [18]. Ethane (C_2H_6) exhibits similar phase behavior, differing by only a few degrees. In an earlier study [7], it was suggested that Titan's oceans would likely consist predominantly of ethane (75%), with methane comprising about 25%, and minor contributions from dissolved nitrogen (N_2), as inferred from Voyager data indicating an unsaturated nitrogen adiabat above a solid surface.

The second-largest sea on Titan, after Kraken Mare (68° N, 310° W; see **Figures 2** and **3**), is Ligeia Mare (78° N, 249° W). The first bathymetric cross-section of Ligeia Mare was obtained by Mastrogiuseppe et al. [19] using Cassini fly-by T91 altimetry data, revealing a maximum depth approaching 160 m (see **Figure 4**). Remarkably, the quality of the radar signal suggests that the lake is extremely transparent, consistent with a composition dominated by methane and ethane and a paucity of suspended particulates.

Evaporation rates on Titan have been estimated at approximately 20 mm per week at the poles [20] and about 1 mm per week at the equator [21]. Observations by Barnes et al. [22], using Cassini/VIMS during the T85 fly-by, revealed a specular solar reflection over Kivu Lacus (87.4° N, 241.1° E). Their radiative transfer modeling

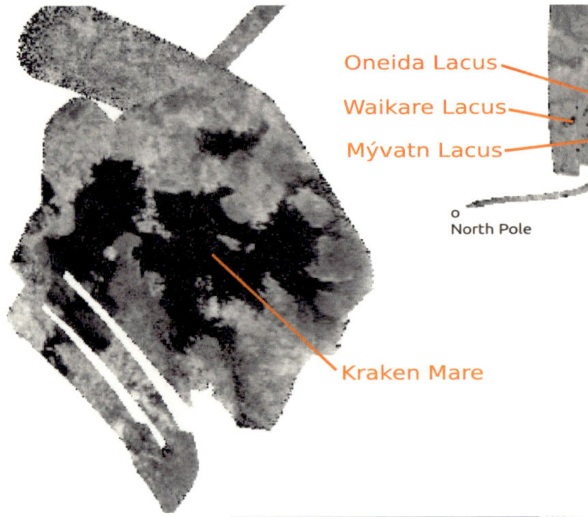

Figure 2.
Kraken Mare & small lakes (fly-by T91 from pds-imaging.jpl.nasa.gov).

Figure 3.
Ligeia Mare at the upper central part of the map of the main Titan oceans, USGS, public domain.

indicated that the total haze column abundance in this region is 20% higher than the Huygens probe's equatorial measurement, strongly suggesting the presence of atmospheric feedback processes associated with polar lakes.

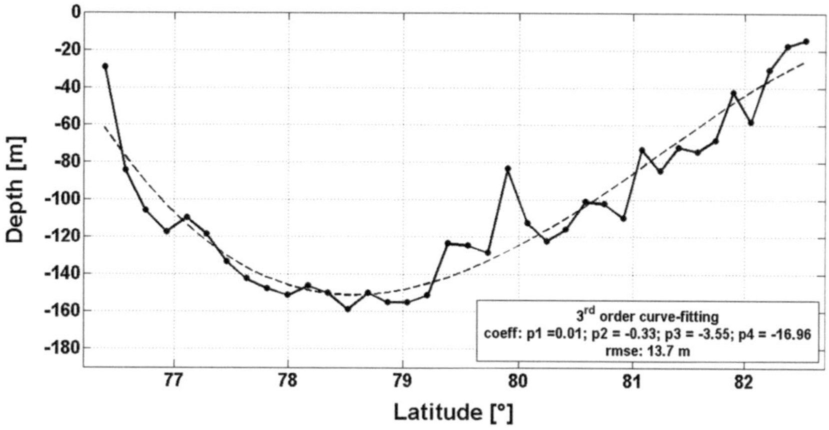

Figure 4.
Ligeia Mare Bathymetry cross-section (fly-by T91), from the study by Mastrogiuseppe et al. [19].

In addition, Savage et al. [23] reported that approximately 13% of Titan's global surface is covered by linear dune fields, primarily located within 30 degrees of the equator. These mature dunes, averaging 1.3 km in width and 2.7 km in crest spacing, become more prevalent toward the equator, where surface moisture is lower than at higher latitudes. Their composition likely consists of complex, fine-grained hydrocarbon molecules, possibly derived from atmospheric photolysis. Saltation of these grains requires wind velocities exceeding 1 m s^{-1}, a threshold typically achieved only during equinox conditions. For most of the Titanian year, wind speeds are expected to be lower, and the prevailing wind directions at these lower velocities remain poorly constrained. Assuming the regolith or lake/mare surfaces are saturated with hydrocarbons in the liquid phase, evaporation rates will be strongly influenced by aerodynamic factors, including wind speed, surface roughness, and the degree of atmospheric saturation with respect to the relevant hydrocarbons.

4. Rainfall

Early work by Lunine et al. [7] proposed several photolytic pathways for methane (CH_4) in Titan's mesosphere at wavelengths shorter than 1,450 Å:

$$2CH_4 \rightarrow C_2H_4 + 2H + H_2 \quad (5)$$

$$2CH_4 \rightarrow C_2H_2 + 3H_2 (or\ H_2 + 4H) \quad (6)$$

In the stratosphere, photolysis of methane at wavelengths between 1,450 Å and 2,000 Å can proceed via:

$$2CH_4 \rightarrow C_2H_6 + H_2 \quad (7)$$

$$2CH_4 \rightarrow C_2H_6 + 2H \quad (8)$$

In these reactions, acetylene (C_2H_2) acts as a catalyst, while atmospheric ethane (C_2H_6) is ultimately lost through condensation [7].

Titan's Inter-Tropical Convection Zone (ITCZ; see **Figure 5**) migrates from pole to pole [24], a consequence of the moon's slow rotation, resulting in pronounced seasonal variations at the surface, as documented by Cassini's SAR and VIMS instruments [25]. Further hypotheses suggest the potential for tropical cyclones above polar lakes during the northern summer solstice, when increased sea surface temperatures and enhanced near-surface air vorticity – –driven by seasonal convergence and equatorial wave activity – –may favor cyclone formation.

Additional studies, such as by Dalba et al. [6], indicate that rainfall on Titan can evaporate, sublimate, or infiltrate within timescales of months, and is sometimes associated with fluvial features, though not necessarily with their formation or modification [26]. Estimated that polar rainfall can reach 2,000–4,000 mm per Titan year (covering approximately 15% of the surface), resulting in a global mean of 300–600 mm per Titan year, values consistent with global radiation budgets. Prevailing winds are predominantly eastward [27], with seasonal shifts giving rise to Aeolian dune fields concentrated within ±30 degrees of the equator, regions where precipitation is less frequent. Rainstorm intensities on Titan can reach several hundred millimeters per hour [28, 29], highlighting the potential for episodic but intense surface modification.

Observations of low-latitude cloud systems have revealed extensive surface changes [15], suggesting that seasonal precipitation events are responsible for carving dry channels in equatorial regions – paralleling the geomorphic dominance of fluvial activity in many terrestrial arid environments. Supporting this, Beatty [30] found evidence that equatorial areas on Titan have experienced greater erosion than polar regions.

5. Run-off and erosion processes

The absence of a dedicated orbiter for Titan has limited the availability of high-resolution, continuous surface data. To date, the Cassini mission has provided only

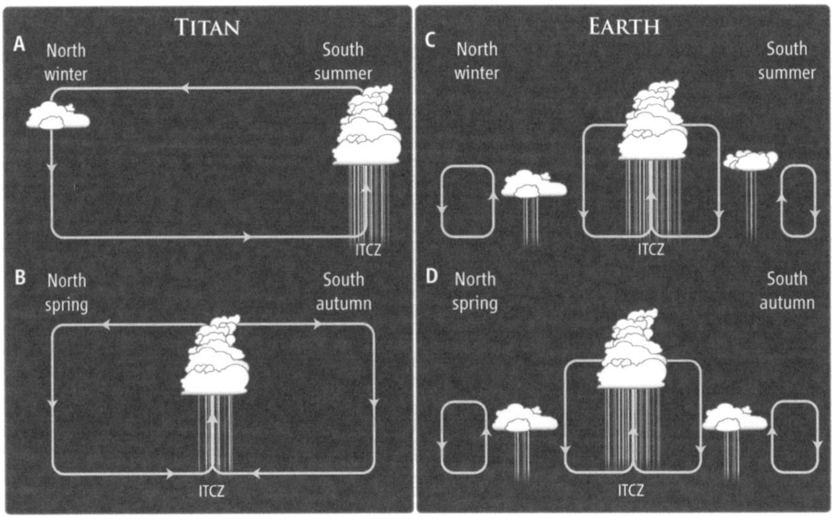

Figure 5.
Inter-Tropical Convection Zone on Titan and Earth, from the study by Tokano [24].

fly-by observations, and while radar altimeter data may eventually enable the construction of a global elevation model for Titan's surface, the expected resolution – ranging from several hundred to a thousand meters – remains insufficient for detailed studies of run-off processes [31]. Notably, run-off on Titan may be highly concentrated in sediments [32], an intriguing possibility given the remarkable transparency of Ligeia Mare, as observed by [19].

Erosional processes on Titan appear to be primarily mechanical rather than chemical in nature [18]. Morphological analyses of landscape evolution indicate that the northern lakes are predominantly shaped by fluvial erosion, with less than 9% of their initial area being eroded over time. Furthermore, one of the two drainage shape parameters assessed was found to be sensitive only to less-developed networks. The estimated timescale for the development of these features is approximately 45,000 years, aligning with previous estimates of Titan's surface age.

Additional research by Malaska et al. [33] examined bright lineated plains, streak-like plains, and linear dunes to infer the primary sediment transport pathways across Titan. Their results suggest that, in both hemispheres, materials from 0° to 35° latitude are transported poleward toward a belt centered around 35° latitude, while materials from 60° to 35° latitude are transported equatorward toward the same belt. Aeolian transport was identified as the dominant mechanism in the equatorial and mid-latitude zones. Interestingly, only plains units are observed at approximately 35° latitude, leading to the hypothesis that dune materials may be converted or modified into plains units, or that sediment transport at dune margins is limited by supply.

A study by Neish et al. [34] of five confirmed craters larger than 20 km in diameter revealed a uniform distribution of crater relative depths, suggesting an infilling process that progresses linearly over time. This observation is consistent with findings from Mars, where Aeolian infill also varies linearly with time [35], while fluvial modification rates diminish over time. These results raise important questions regarding the temporal distribution and frequency of rainfall-runoff events on Titan.

At the Huygens landing site (10.2° S, 192.4° W), acoustic measurements indicated the presence of sorted, interlocking grains that are smooth at the centimeter scale, consistent with fluvial sorting, or alternatively, the infilling of small particles among larger ones, analogous to terrestrial playas [36]. Methane may also act as an interstitial liquid or exist in small pools within the substrate. Penetrometer data from Huygens suggest that sand particles approximately 2 mm in diameter are overlain by a potential 7 mm layer of ice-like material [37]. The observed decrease in resistance with depth implies incremental moistening of the substrate. Additionally, ice-water pebbles have been identified in pluvial and fluvial geomorphological contexts in the southwest region of Xanadu (SW of 15° S, 100° W), as well as at the Huygens landing site [38]. In some cases, only large flood events could account for the observed mass movement of these materials.

6. Conclusion

Titan stands out as a unique planetary body in the solar system, exhibiting a dense, nitrogen-rich atmosphere and a dynamic surface shaped by complex hydrocarbon cycles. The combination of Cassini–Huygens observations and recent advances in ground- and space-based spectroscopy has revealed an active methane-based meteorology, with surface and atmospheric processes that parallel, yet fundamentally differ from those on Earth. Titan's thick haze layers, driven by the photochemistry of methane and nitrogen, not only obscure the surface but also play a crucial role in the moon's thermal structure and seasonal variations.

The detection of a diverse suite of organic molecules (including ethane, ethyl cyanide, vinyl cyanide, cyclopropenylidene, propadiene, and tricarbon species) demonstrates the richness of Titan's atmospheric chemistry and its potential as a natural laboratory for prebiotic processes. Seasonal and spatial variability in cloud formation, rainfall, and surface albedo highlights the importance of Titan's energy balance and the interplay between atmospheric and surface reservoirs. The evidence for active evaporation, precipitation, and run-off, as well as sediment transport and fluvial erosion, underscores the ongoing reshaping of Titan's landscape by processes analogous to, but distinct from, terrestrial hydrology.

Despite these advances, key questions remain regarding the sources and sinks of methane, the detailed mechanisms of organic synthesis, and the long-term evolution of Titan's climate. The limitations of current data, stemming from the absence of a dedicated orbiter and the challenges of remote sensing through Titan's opaque atmosphere, underscore the need for future missions equipped to probe both the surface and atmosphere with greater spatial and temporal resolution. Continued exploration of Titan promises not only to deepen our understanding of this intriguing moon but also to shed light on the processes that may have shaped early Earth and other planetary bodies with thick, nitrogen-rich atmospheres.

Abbreviations

IRTF-TEXES	Infrared Telescope Facility-Texas Echelon Cross Echelle Spectrograph
ITCZ	Inter Tropical Convection Zone
JWST	James Webb Space Telescope
LW	Long Wave
MIRI	Mid-Infrared Instrument onboard JWST
SAR	Synthetic Aperture Radar
SW	Short Wave
VIMS	Visual and Infrared Mapping Spectrometer

Author details

Yann-Henri Chemin
Seilio Douar, Plumergat, France

*Address all correspondence to: dr.yann.chemin@gmail.com

References

[1] Hörst SM. Titan's atmosphere and climate. Journal of Geophysical Research: Planets. 2017;**122**:432–482

[2] Hayes AG. Hydrocarbon lakes on Titan. 2012. Video of a talk given at SETI. Available from: https://www.youtube.com/watch?v=e9sfNegkpPE [Accessed: 2025-July-09]

[3] Nixon CA, Teanby NA, Irwin PGJ, et al. The atmosphere of Titan in late northern summer from JWST and Keck observations. Nature Astronomy. 2025;**9**:123–135

[4] Mousis O, Lunine JI, Pasek M, Cordier D, Waite JH, Mandt KE, Lewis WS, Nguyen M-J. A primordial origin for the atmospheric methane of Saturn's moon Titan. Icarus. 2009;**204**(2):749–751

[5] Mandt KE, Waite JH, Teolis B, Magee BA, Bell J, Westlake JH, Nixon CA, Mousis O, Lunine JI. The 12C/13C ratio on Titan from Cassini INMS measurements and implications for the evolution of methane. The Astrophysical Journal. 2012;**749**(2):160

[6] Dalba PA, Buratti BJ, Brown RH, Barnes JW, Baines KH, Sotin C, Clark RN, Lawrence KJ, Nicholson PD. Cassini VIMS observations show ethane is present in Titan's rainfall. The Astrophysical Journal Letters. 2012;**761**(2):L24

[7] Lunine JI, Stevenson DJ, Yung YL. Ethane ocean on Titan. Science. 1983;**222**(4629):1229–1230

[8] Cordiner MA, Palmer MY, Nixon CA, et al. Ethyl cyanide on Titan: Spectroscopic detection and mapping using ALMA. The Astrophysical Journal Letters. 2015;**800**:L14

[9] Nixon CA, Thelen AE, Cordiner MA, Kisiel Z, Charnley SB. Detection of cyclopropenylidene (c-C3H2) on Titan with ALMA. The Astronomical Journal. 2020;**160**:181

[10] Lombardo NA, Nixon CA, Irwin PGJ, et al. Detection of propadiene on Titan. The Astrophysical Journal Letters. 2019;**881**:L33

[11] Riança-Silva R, Machado P, Martins Z, Lellouch E, Loison J-C, Dobrijevic M, Dias JA, Ribeiro J. A study of very high resolution visible spectra of Titan: Line characterisation in visible CH_4 bands and the search for C_3. Planetary and Space Science. 2024;**240**:106228

[12] Burr DM, Perron JT, Lamb MP, Irwin RP, Collins GC, Howard AD, Sklar LS, Moore JM, Ádámkovics M, Baker VR. Fluvial features on Titan: Insights from morphology and modeling. Geological Society of America Bulletin. 2013;**125**(3–4):299–321

[13] Flasar FM. Oceans on Titan? Science. 1983;**221**(4605):55–57

[14] McKay CP, Pollack JB, Courtin R. The greenhouse and antigreenhouse effects on Titan. Science. 1991;**253**(5024):1118–1121

[15] Turtle EP, Perry JE, Hayes AG, Lorenz RD, Barnes JW, McEwen AS, West RA, Del Genio AD, Barbara JM, Lunine JI, Schaller EL, Ray TL, Lopes RMC, Stofan ER. Rapid and extensive surface changes near Titan's equator: Evidence of april showers. Science. 2011;**331**(6023):1414–1417

[16] Gillespie A, Rokugawa S, Matsunaga T, Cothern JS, Hook S, Kahle AB. A temperature and emissivity separation algorithm for advanced spaceborne thermal emission and reflection radiometer (ASTER) images. IEEE Transactions on Geoscience and Remote Sensing. 1998;**36**(4):1113–1126

[17] Waite JH, Bell J, Lorenz R, Achterberg R, Flasar FM. A model of variability in Titan's atmospheric structure. Planetary and Space Science. 2013;**86**(0):45–56

[18] Black BA, Perron JT, Burr DM, Drummond SA. Estimating erosional exhumation on Titan from drainage network morphology. Journal of Geophysical Research: Planets (1991–2012). 2012;**117**(E8):1–17

[19] Mastrogiuseppe M, Poggiali V, Hayes A, Lorenz R, Lunine J, Picardi G, Seu R, Flamini E, Mitri G, Notarnicola C. The bathymetry of a Titan sea. Geophysical Research Letters. 2014;**41**:1432–1437

[20] Hayes AG, Aharonson O, Lunine JI, Kirk RL, Zebker HA, Wye LC, Lorenz RD, Turtle EP, Paillou P, Mitri G, Wall SD, Stofan ER, Mitchell KL, Elachi C. Transient surface liquid in Titan's polar regions from Cassini. Icarus. 2011;**211**(1):655–671

[21] Mitchell JL. The drying of Titan's dunes: Titan's methane hydrology and its impact on atmospheric circulation. Journal of Geophysical Research. 2008;**113**(E8):E08015

[22] Barnes JW, Clark RN, Sotin C, Ádámkovics M, Appéré T, Rodriguez S, Soderblom JM, Brown RH, Buratti BJ, Baines KH, LeMouélic S, Nicholson PD. A transmission spectrum of Titan's North Polar atmosphere from a specular reflection of the Sun. The Astrophysical Journal. 2013;**777**(2):161

[23] Savage CJ, Radebaugh J, Christiansen EH, Lorenz RD. Implications of dune pattern analysis for Titan's surface history. Icarus. 2014;**230**:180–190

[24] Tokano T. Precipitation climatology on Titan. Science. 2011;**331**:1393–1394

[25] Tokano T. Are tropical cyclones possible over Titan's polar seas? Icarus. 2013;**223**(2):766–774

[26] Lorenz RD, Turtle EP. How often does it rain on Titan? Lunar and Planetary Institute Science Conference Abstracts. 2012;**43**:2472

[27] Radebaugh J, Lorenz RD, Lunine JI, Wall SD, Boubin G, Reffet E, Kirk RL, Lopes RM, Stofan ER, Soderblom L, Allison M, Janssen M, Paillou P, Callahan P, Spencer C, Cassini radar team. Dunes on Titan observed by Cassini RADAR. Icarus. 2008;**194** (2):690–703

[28] Hueso R, Sànchez-Lavega A. Methane storms on Saturn's moon Titan. Nature. 2006;**442**:428

[29] Barth EL, Rafkin SCR. Convective cloud heights as a diagnostic for methane environment on Titan. Icarus. 2010;**206**:467

[30] Beatty JK. Titan's exterior offers stiff resistance. Sky & Telescope. 2013;**126**(6):16

[31] Lorenz RD, Stiles BW, Aharonson O, Lucas A, Hayes AG, Kirk RL, Zebker HA, Turtle EP, Neish CD, Stofan ER, Barnes JW. A global topographic map of Titan. Icarus. 2013;**225**(1):367–377

[32] Burr DM, Emery JP, Lorenz RD, Collins GC, Carling PA. Sediment transport by liquid surficial flow: Application to Titan. Icarus. 2006;**181**(1):235–242

[33] Malaska MJ, Lopes RMC, Hayes AG, et al. Material transport map of Titan: The fate of dunes. Icarus. 2016;**270**:183–196

[34] Neish CD, Kirk RL, Lorenz RD, Bray VJ, Schenk P, Stiles B, Turtle E, Cassini RADAR Team. Crater topography on Titan: Implications for landscape evolution. Lunar and Planetary Institute Science Conference Abstracts. 2012;**43**:2412

[35] Forsberg-Taylor NK, Howard AD, Craddock RA. Crater degradation in the Martian highlands: Morphometric analysis of the Sinus Sabaeus region and simulation modeling suggest fluvial processes. Journal of Geophysical Research: Planets (1991–2012). 2004;**109**(E5):1–12

[36] Towner MC, Garry JRC, Lorenz RD, Hagermann A, Hathi B, Svedhem H, Clark BC, Leese MR, Zarnecki JC. Physical properties of Titan's surface at the huygens landing site from the surface science package acoustic Properties Sensor (API-S). Icarus. 2006;**185**(2):457–465

[37] Atkinson KR, Zarnecki JC, Towner MC, Ringrose TJ, Hagermann A, Ball AJ, Leese MR, Kargl G, Paton MD, Lorenz RD, Green SF. Penetrometry of granular and moist planetary surface materials: Application to the Huygens landing site on Titan. Icarus. 2010;**210**(2):843–851

[38] Le Gall A, Janssen MA, Paillou P, Lorenz RD, Wall SD. Cassini Radar Team, the. Radar-bright channels on Titan. Icarus. 2010;**207**(2):948–958

Chapter 3

Variability of Polarization on the Surface of Jupiter's and Saturn's Moons

Nikolai Kiselev and Marina Shcherbina

Abstract

We present the disk-integrated longitude-polarization functions of Jupiter's and Saturn's satellites. The observed degree of linear polarization, $P(obs)$, of atmosphereless/airless solar system bodies (ASSBs) – asteroids and planetary satellites – depends on many quantities and observation circumstances: $P(obs) = P$ (phase angle, planetocentric longitude, latitude, time, and wavelength). The dependence of polarization, $P(\alpha)$, on the phase angle is most sensitive to the properties of the regolith surfaces. However, in some cases, it is possible to detect polarization changes over the body surface, the so-called longitudinal dependence of polarization (LDP). Accurate measurements of $P(obs)$ at different phase angles and longitudes have made it possible to establish the LDP of Jupiter's satellites (Europa, Io, Ganymede, and Callisto) and Saturn's satellites (Iapetus and Rhea). This may be evidence of optical/structural inhomogeneities on the surface of the bodies. A long-term change in the longitude dependence of Io's polarization may indicate a change in global and/or local volcanic activity on the satellite's surface.

Keywords: polarization, phase angle dependence of polarization, longitudinal dependence of polarization, planetary satellites, properties of surface

1. Introduction

Scattered light produced when unpolarized radiation interacts with various particles and media exhibits a number of properties, one of the most important being linear polarization. The observed degree of linear polarization, P_{obs}, for airless/atmosphereless solar system bodies (ASSBs) – such as asteroids, planetary satellites, and comets – generally depends on multiple quantities and observing conditions. This can be written as

$$\mathrm{P(obs) = P(\alpha, L, t, \lambda)} \tag{1}$$

where α is the phase angle; L is the planetocentric longitude (for planetary satellites and asteroids); t is the time of observation (for non-stationary objects such as comets or active asteroids); and λ is the wavelength.

The phase-angle dependence of polarization, $P(\alpha)$ (the polarization phase curve [PPC]), is especially sensitive to the properties of regolith-covered surfaces of ASSBs. In addition, in a number of cases, it is possible to detect variations in polarization across a body's surface – the so-called longitudinal dependence of polarization (LDP) – which can indicate optical/structural inhomogeneities on the surface. In this review, we present examples of LDP detections for several satellites of Jupiter and Saturn, for which the most accurate PPCs to date have been obtained.

2. Phase dependence of the polarization of scattered radiation

Before discussing spatial variations in the polarization degree across planetary satellites, we briefly recall the key parameters of the polarization phase dependence that must be accounted for when searching for an LDP.

Electromagnetic waves (EM waves) are transverse to the direction of propagation. For thermal radiation sources, in the absence of magnetic fields, the oscillations of the electric vectors in the image plane have no preferred direction. However, the interaction of transverse waves with a surface or medium introduces an asymmetry in the distribution of electric-vector orientations in the image plane. The degree of linear polarization, P, quantifies this asymmetry:

$$P = \frac{I_{max} - I_{min}}{I_{max} + I_{min}} = \frac{I_p}{I} \qquad (2)$$

where $I_{\max}$ and $I_{\min}$ are the maximum and minimum intensities (squared EM amplitudes), and I_{p} and I are the polarized and total intensities, respectively. By definition, $0 \le P \le 1$. The angle at which the intensity is maximal in the image plane, θ, is the position angle of the polarization plane. The quantities P and θ are usually specified in a chosen reference frame. In **Figure 1**, P and θ are shown with respect to the equatorial system, in which the position angle is measured from celestial north through east (counterclockwise). The polarization parameters P and θ are commonly expressed via the Stokes parameters Q and U [1]:

$$Q = P\cos 2\theta, U = P\sin 2\theta \qquad (3)$$

In planetary astrophysics, the scattering plane is adopted as the reference frame. If φ is the position angle of the scattering plane, the transformation to this frame is

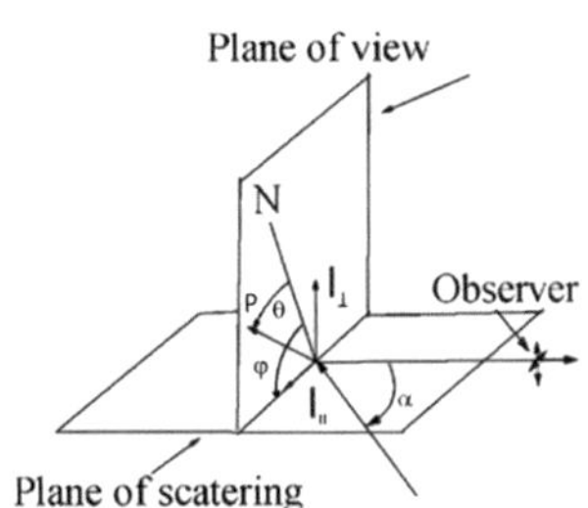

Figure 1.
Schematic illustrating the principal characteristics of polarized scattered radiation. P *and* θ *are the degree and position angle of linear polarization in the equatorial frame.* A *is the phase angle;* φ *is the position angle of the scattering plane.* $I_{||}$ *and* $I_{\perp}$ *are the intensities parallel and perpendicular to the scattering plane.*

$$\theta_r = \theta - (\varphi \pm 90°), P_r \equiv Q_r = P\cos 2\theta_r, U = P\sin 2\theta_r \tag{4}$$

Because the perpendicular to the scattering plane can be chosen in two ways, the sign in (φ ±90°) is selected so that its value does not exceed 180°.

In the scattering-plane frame, the (signed) polarization can also be written as.

$$P_r = (I_\perp - I_{II})/(I_\perp + I_{II}) = (I_\perp - I_{II})/I \tag{5}$$

where $I_\perp$ and $I_{||}$ are the intensities of the scattered light with electric vectors oscillating perpendicular and parallel to the scattering plane, respectively (see **Figure 1**). As noted above, P is non-negative by definition; however, according to Eq. (5), if $I_\perp < I_{||}$, the quantity P_r is formally negative. Physically, this means that the electric vectors oscillate predominantly in the scattering plane. This behavior is characteristic of most airless bodies (planetary satellites, asteroids) and of comets observed near opposition, at phase angles $\alpha \le \alpha_{inv} \approx 9° - 30°$.

3. Instrumentation and methodology for determining the LDP in planetary satellites

In 2018, two aperture two-channel polarimeters, "POLSHAKH," were built at the Crimean Astrophysical Observatory (CrAO) [2]. They are used for polarimetric observations on the 2.6-m CrAO telescope and the 2-m telescope of the Peak Terskol Observatory. The instruments are designed to measure the parameters of linear and circular polarization both in broadband UBVRI filters and in dedicated narrowband filters for various celestial objects. The determination of polarization parameters is based on measuring the intensity of light transmitted through a continuously rotating (≈30 Hz) half-wave/quarter-wave retarder, followed by splitting the beam with a Wollaston prism into two orthogonally polarized components.
A synchronous-detection scheme is employed, that is, photon counts are accumulated in computer memory cells in step with the retarder's rotation angle. As a result, the observational outcomes are practically insensitive to short-term atmospheric conditions. The instrumental polarization of both polarimeters does not exceed 0.04% and varies only weakly from season to season, which – after removal – allows us to investigate subtle polarimetric effects of celestial bodies.

As noted above, after correcting for instrumental effects and transforming to the scattering-plane reference system, the observed degree of linear polarization, P_r, depends on the observing geometry, namely on the phase angle, α, and other factors. Observations over different phase angles yield the PPC, $P_r(\alpha)$.

Because most satellites are tidally locked, they always present the same face to their parent planet. Consequently, as viewed from Earth, different hemispheres are seen at eastern versus western elongations. The planetocentric longitude L of a satellite is defined as the angle measured from the point of superior geocentric conjunction, increasing counterclockwise as seen from the north celestial pole. Over $0° \le L \le 180°$, the leading hemisphere (facing the direction of orbital motion) is observed; whereas, for $180° < L < 360°$, the trailing hemisphere is seen. If the physical/structural properties of the surface vary with location, the observed linear polarization P_r may also depend on longitude L.

Because of the satellites' orbital inclinations relative to the line of sight to Earth, the observed surface regions differ not only in longitude but also in latitude. Therefore, both longitudinal and latitudinal polarization effects are possible. We adopt as L and β the apparent planetographic longitude and latitude of the disk center, taken from JPL ephemerides [3].

Observing geometry allows Jupiter's satellites to be studied over phase angles 0° -12°, and Saturn's satellites over 0° - 6°. To reliably disentangle the phase and longitudinal dependences of polarization, measurements must therefore be obtained over a wide range of longitudes and latitudes and at multiple phase angles, which requires a substantial observational data set. In our material, the maximum latitude reached for Rhea was $\beta = 22°$; for the other satellites, β was limited to only a few degrees. Hence, latitudinal polarization effects are neglected in this work.

Because the dependence on phase angle α is the strongest, the relation P_r versus α is customarily termed the phase dependence of the polarization degree, although for some objects it may implicitly depend on longitude, that is, $P_r(\alpha, L)$. In the range $0 \le \alpha \le \alpha_{inv}$, the phase curve exhibits a negative polarization branch (NPB), whose schematic form is shown in **Figure 2**.

3.1 General procedure for extracting the LDP from observations

To first order, the observed degree of linear polarization $P_r(\alpha, L)$ can be represented as a superposition of two dependencies:

$$P_r(\alpha, L) = P_r(\alpha) + A(\alpha)P_r(L) \quad (6)$$

where $P_r(\alpha)$ is the PPC (PDP), free of longitudinal effects; $P_r(L)$ is the dependence of polarization on orbital longitude at a given phase angle; and $A(\alpha)$ is the amplitude of the LDP, which may vary with phase. Over a narrow range of phase angles, the LDP can be treated as constant. Over the broader range $0 \le \alpha \le \alpha_{inv}$, $P_r(\alpha, L)$ is not linear in α, especially near α_{min}. Therefore, setting $P_r(L) = 0$

In Eq. (6), we selected a phase-angle interval $\Delta\alpha$ within which the values of $P_r(\alpha, L)$ can be approximated by a straight line. Then, to first order and separately for each hemisphere of a satellite, one can write.

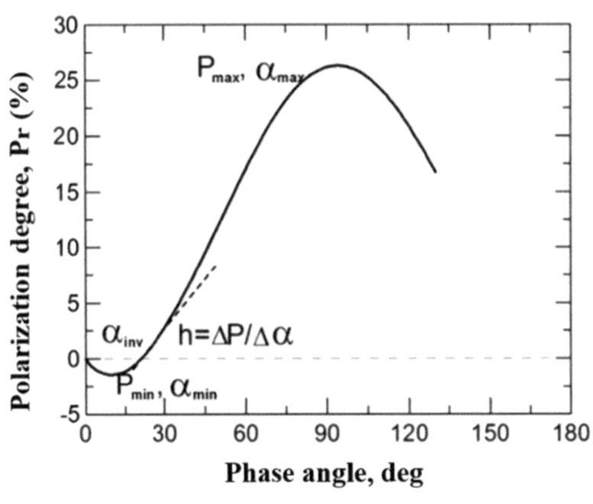

Figure 2.
Characteristic shape of the observed polarization dependence for airless solar system bodies and comets, and its parameters over a broad range of phase angles. Parameters of the negative branch include P_{min} *and* α_{min} *(the minimum polarization and the corresponding phase angle), the inversion angle* α_{inv}, *and the polarimetric slope at the inversion angle* $h = \Delta P/\Delta\alpha$ *(in percentage per degree).*

$$P_r(\alpha) = aP_r(\alpha, L) + b \tag{7}$$

where a and b are the coefficients of the linear fit, obtained by least squares. Consequently – accounting for the fitted slope a – the longitudinal dependences for the leading and trailing hemispheres were evaluated at the midpoint of the chosen. $\Delta\alpha$ as

$$P_r(\mathrm{L}) = P_r(\alpha, \mathrm{L}) - aP_r(\alpha, \mathrm{L})(\alpha - \Delta\alpha/2) \tag{8}$$

Below, we present the polarization measurements of planetary satellites obtained with these polarimeters over 2018–2025.

4. Results on the LDP for selected planetary satellites

4.1 Jupiter's satellites

4.1.1 Europa

Europa's surface is covered by an ice shell with a complex network of fractures discovered by the two Voyager spacecraft in 1979, along with a small number of impact craters. Beneath the thick ice crust lies an ocean of saline water. Cryovolcanic activity is observed at the surface, likely sourced from the subsurface saline ocean [4–6]. These active phenomena may modify the surface properties. The icy surface makes Europa one of the most reflective objects in the solar system: its disk-averaged geometric albedo in the V band is 0.68. The light scattered by Europa exhibits a steep, nonlinear increase in brightness toward opposition at phase angles $\alpha \leq 2°$ for both the leading and trailing hemispheres. In the visible, the leading hemisphere is brighter than the trailing one [7], generally attributed to the preferential bombardment of the trailing side by charged ions, which darkens the surface.

Extensive polarimetric observations of Europa were conducted by Dollfus [8] and by Chigladze [9], but the most accurate PPCs in the UBVRI bands were obtained only recently by Kiselev et al. [10]. That study noted: "There is a slight difference (on average ~0.02%) in the polarization degree between Europa's two hemispheres only in the R band: the polarization is systematically lower for the leading hemisphere. Most likely, if a longitudinal effect exists, it is small, but this requires verification with additional observations." Motivated by this, we carried out additional observations of Europa, allowing us to refine the shape of the LDP. The final phase dependences of Europa's polarization degree in the VRI bands are shown in **Figure 3** (left panels). The right panels present the longitudinal dependences of polarization normalized to a phase angle of 9° after accounting for the PPC of each hemisphere over α = 5° - 12°. From the right panels (positive polarization branch), it is seen that in the V and R bands, the polarization of the leading hemisphere is systematically lower than that of the trailing hemisphere by ≈0.03% and ≈0.02%, respectively. In the left panels (NPB), the polarization of the leading hemisphere is systematically smaller in absolute value than that of the trailing hemisphere. Thus, the character of the LDP changes between the negative and positive branches of the phase curve, consistent with a phase-angle–dependent spectral behavior of the polarization. No longitudinal dependence is detected in the I band.

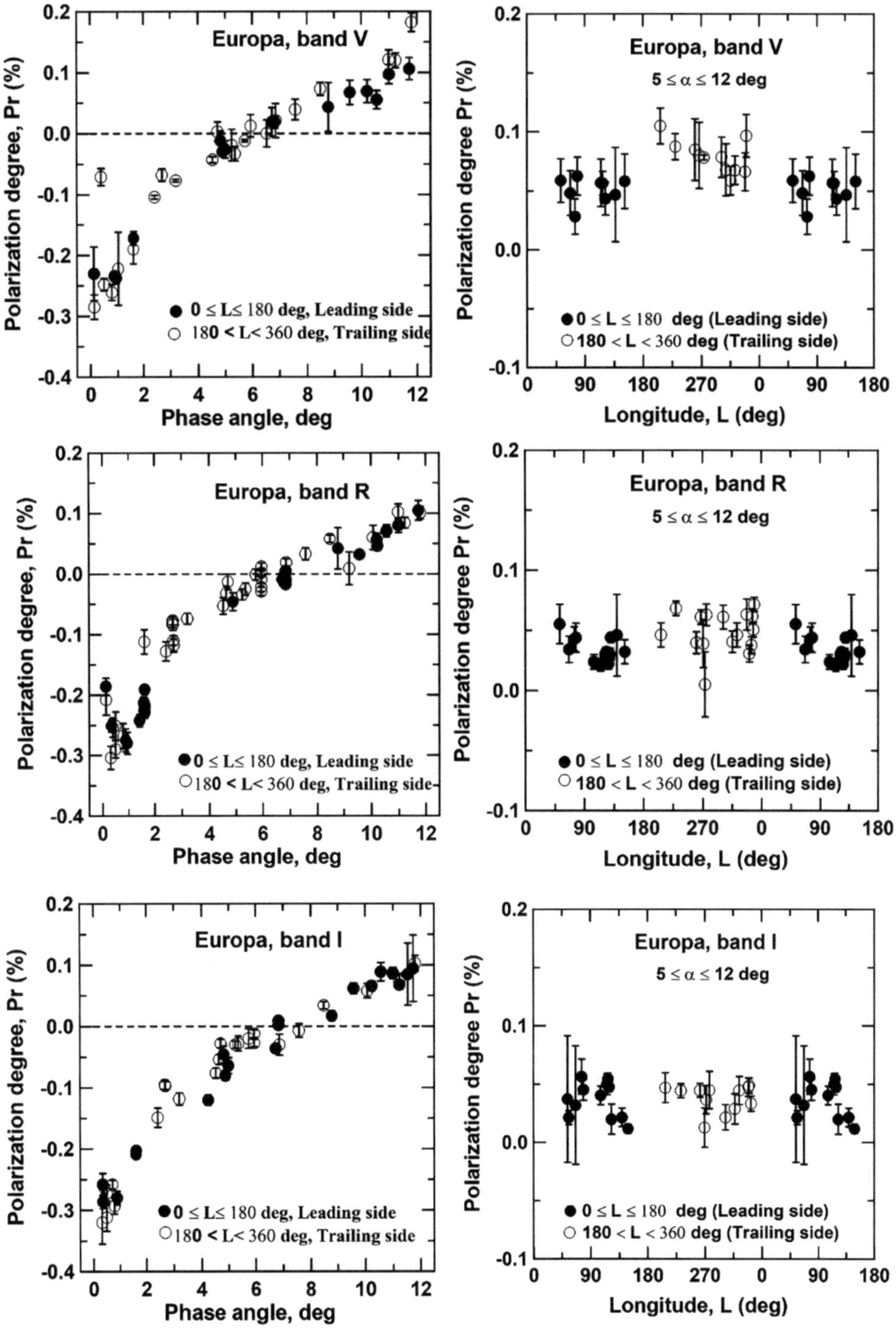

Figure 3.
Left panels: Phase dependences of Europa's polarization degree in the VRI bands. Filled circles – data for the leading hemispheres, 0° ≤ L ≤180°; open circles – data for the trailing hemispheres, 180° < L < 360°. We combine results from Kiselev et al. [10] with additional observations obtained in the period 2022–2023. Right panels: Longitudinal dependences of Europa's polarization after accounting for linear approximations to the observed PPC of each hemisphere over α = 5°–12°.

4.1.2 Io

Io is the most volcanically active object in the solar system; its disk-integrated geometric albedo in the V band is ≈0.63. Its compositionally heterogeneous surface is formed by lava flows from several hundred active volcanoes. Bright flows dominate over dark ones and are interpreted as sulfurous, whereas the dark areas are primarily silicate crust overlain by pyroclastics and mixed-composition flows. This structure produces a pronounced heterogeneity in optical properties [11]. Observations by Galileo [12, 13] showed that, due to the dynamic nature of volcanism on Io, its surface can change over time on a scale of years and even months. Allotropic sulfur compounds give different colors to Io's surface [14]. A similar effect can occur for polarization due to the frost deposits on the surface of Io.

Our polarimetric observations in the period 2019–2024 [15, 16] showed (see **Figure 4**, left panel) that Io's negative polarization decreases sharply in the phase angle range from 0° to about 2°, and then it increases almost linearly to ~12°, reaching a value of about -0.15%. The extrapolated inversion angle in the V and R bands is about 26° ± 6°. For comparison, the right panel of **Figure 4** shows Io's PPC previously obtained by Veverka and Burns [17] over the same phase-angle range.

According to Veverka and Burns [17], the large scatter of the degree of polarization, up to about 0.4%–0.5% at phase angles greater than 10°, obtained by Zellner and Gradie, may be caused by the spotty surface of Io. **Figure 4** compares our measurements of Io's polarization with those of Zellner and Gradie [18], obtained in April 1972 and November 1973, and with extensive polarization measurements of Io by Dollfus [8] in April 1972 and October 1973. The amplitude of the variations in P is 0.1%, 0.2%, and 0.45%, respectively. The accuracy of the measurements by Dollfus and Zellner and Gradie was better than 0.03% and was

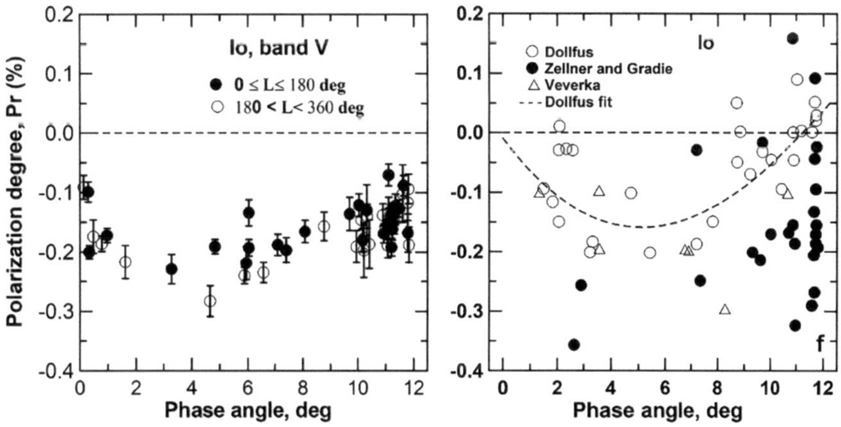

Figure 4.
Left panels – phase-angle dependences of polarization for Io in the V band. The data are taken from literature [15, 16]. Filled circles – data for the leading hemisphere (0° ≤ L ≤ 180°); open circles – data for the trailing hemisphere (180° < L < 360°). The right panel presents data taken from the study by Veverka and Burns [17] (Figure 10.6 of the study), which we adapted to the format of our observations. Filled circles are Io's measurements made by Zellner and Gradie [18] at 0.52 μm. Open circles are data by Dollfus [8] at about 0.5 μm. The observations of Veverka [19] in the clear filter are marked by triangles, and the dashed curve is the fitting polarization curve by Dollfus.

consistent with the accuracy of our measurements. Dollfus believed that the change in Io's polarization was caused by the proximity of Jupiter, which introduced a background of scattered light.

Unfortunately, the results of the measurements by Zellner and Gradie are presented only as a figure in the paper by Veverka and Burns [17]. The lack of the measurement time and Io's longitude precludes verifying Jupiter's contribution to the diffuse light background. However, our observations of Io at different angular distances from Jupiter on the same night of measurements, that is, for close longitudes, showed that variations in the degree of polarization do not exceed the measurement errors. The influence of other potential causes (the positions of the orbiting moon and Io's perihelion), causing polarization variations over a narrow range of phase angles, is difficult to assess. We therefore support the conclusion of Zellner and Gradie that the large scatter of their polarization measurements may be due to local surface properties, which, in turn, are due to Io's volcanic activity.

In literature [12, 13], it has been shown that Io's volcanic activity varies substantially over time. For example, Mekler and Eviatar [20] noted that late 1978 and the first half of 1979 were periods of anomalously high activity; Voyager-era observations likewise indicate the long-lived persistence and variability of volcanic plumes [21]. Taken together with the observed transience of hot spots and plume deposits, these facts indirectly suggest the possibility of seasonal changes in the longitudinal dependence of Io's polarization, driven by long-term variations in both local and global volcanic activity.

Figure 5 shows all observed polarization of Io in the VR bands, in a narrow phase angle range of 10°–12° as a function of longitude L, since we can assume that the phase polarization curves in this range of angles α are constant. The average values of the degree of polarization of both hemispheres are almost the same. It is seen that the data scattering is within about 0.1%, which is close to values of $3^{*}\sigma p$. Therefore, it can mainly be caused by observation errors, and the dependence of polarization on longitude, if any, is insignificant.

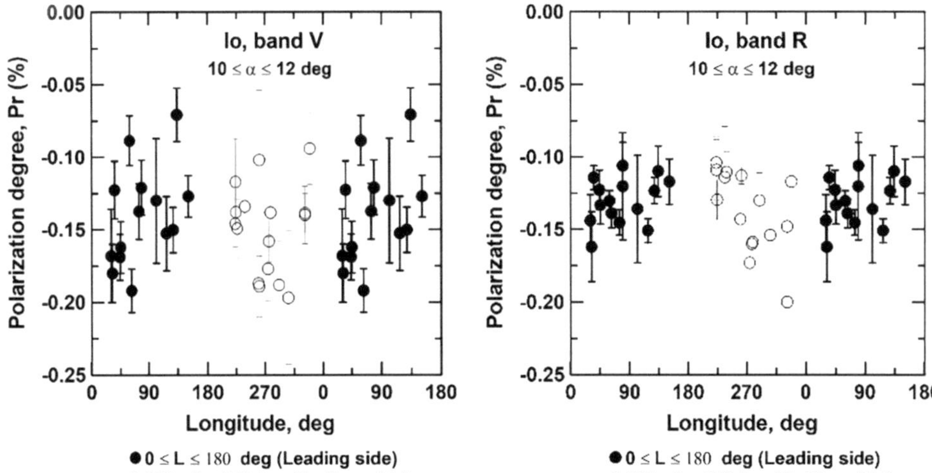

Figure 5.
Longitude dependence of polarization for Io in the V (left slide) and R (right slide) bands in the range of phase angles of 10°–12°.

Figure 6 shows a map of Io's surface compiled from Galileo and Voyager data [22]. It clearly displays color variations and the nonuniform longitudinal distribution of volcanoes, including the largest centers, Pele, Gullan, and Loki. It is quite plausible that the transient activity of these centers can affect the properties of light reflected from Io's surface.

Because of the intrinsically low spatial resolution of the aperture (disk-integrated) polarimetric method – since the measured polarization degree refers to the entire disk (±90° in longitude) at the instant of measurement – the distribution of polarization across the surface cannot be resolved in detail. Thus, even considering the scattering of data, the amplitude of the longitudinal dependence of Io's polarization does not exceed 0.1%. However, the amplitude of Io's LDP differs markedly from the value reported by Zellner and Gradie. Their result was likewise obtained from disk-integrated polarimetric measurements. This contradicts earlier estimates and indicates the variability of the longitudinal effect over large time intervals.

4.1.3 Ganymede

Ganymede is the largest satellite of Jupiter. Unlike Europa's comparatively uniform surface, Ganymede's surface is distinctly heterogeneous, partially ice-covered, and consists of a mixture of two terrain types. The dark, heavily cratered regions occupy roughly one-third of the surface. The brighter areas are younger and are covered with grooves and ridges [8, 19]. The leading hemisphere is brighter than the trailing one. At small phase angles, a brightness opposition effect (BOE) is observed.

New disk-integrated UBVRI polarimetric observations [15] show that Ganymede's NPB is not bimodal: the curve is strongly asymmetric and has a single minimum, similar to Europa. In the V band, we obtained $P_{min} = -0.34\% \pm 0.01\%$ at $\alpha_{min} = 0.52° \pm 0.06°$; the inversion occurs at $\alpha_{inv} = 8.5° \pm 0.2°$. **Figure 7** presents the phase dependences of Ganymede's polarization degree in the VRI bands (left panels) and the longitudinal dependences after accounting for linear fits to the observed PPC of each hemisphere over $\alpha = 5° - 12°$ (right panels).

A clear longitudinal variation is revealed: the (absolute) polarization of the leading side exceeds that of the trailing side. The LDP amplitude increases with wavelength and, in the I band, reaches ~0.15%, with the maximum near L~270°. These features are interpreted as manifestations of genuine differences in regolith properties across the surface.

In sum, the combined photometry/spectroscopy and high-precision polarimetry yield a consistent picture: Ganymede is an icy satellite with a global longitudinal heterogeneity of its regolith. Its NPB has a single sharp minimum at small phase angles, and it exhibits stable longitudinal variations at the level of a few tenths of a percent, explainable by spatial variations in the fractions of ice and dark contaminants, grain size, and surface roughness.

4.1.4 Callisto

Polarimetric and photometric observations of Callisto have been carried out by many authors (see the review by Rosenbush et al. [23] and references therein). As a result, its photometric behavior, as well as the phase and longitudinal dependences of polarization, is well studied.

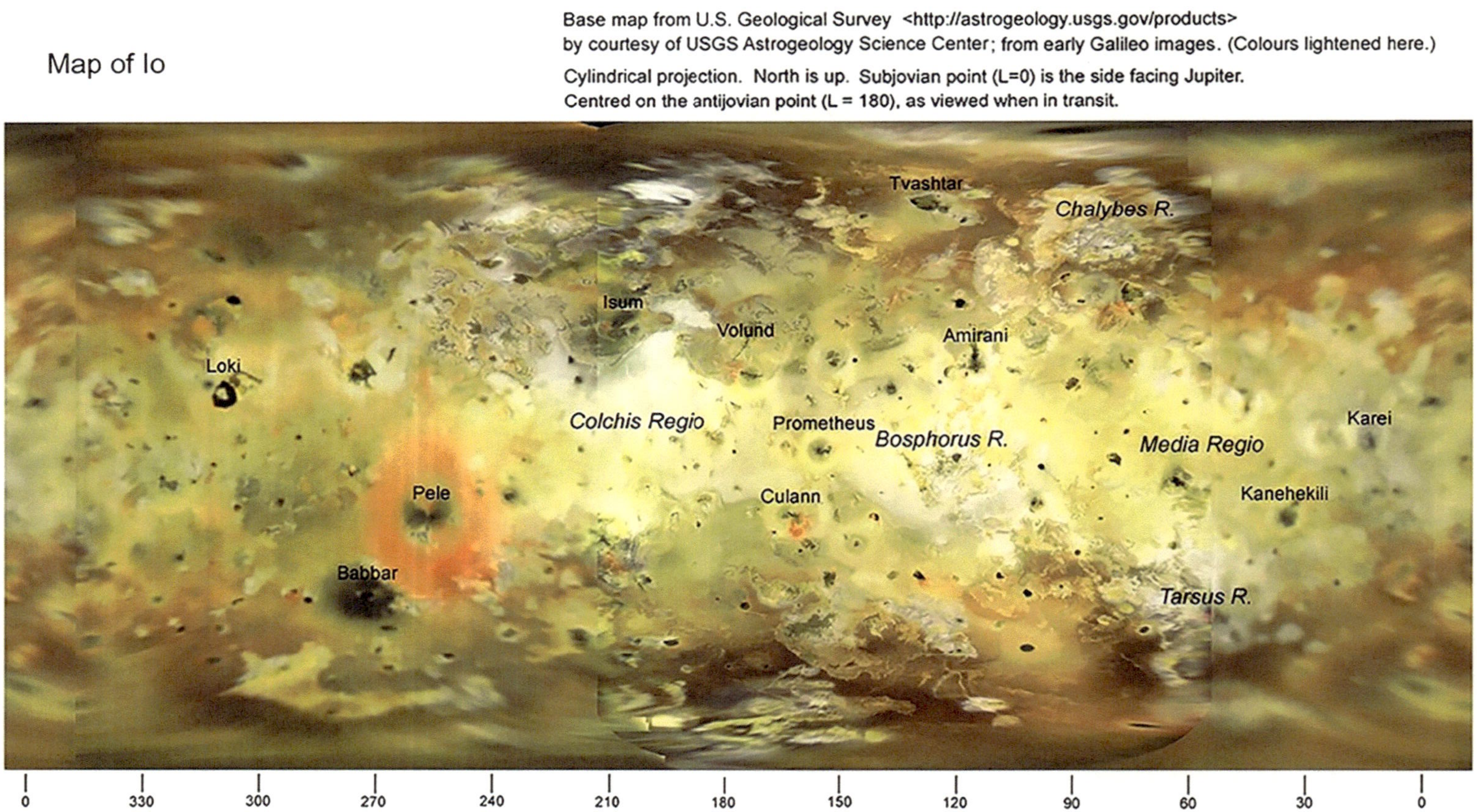

Figure 6.
Cylindrical map of Io (north up), centered on the antijovian longitude (L = 180°). Major volcanic centers are labeled. Base map: USGS Astrogeology Science Center [22], derived from Galileo/Voyager mosaics.

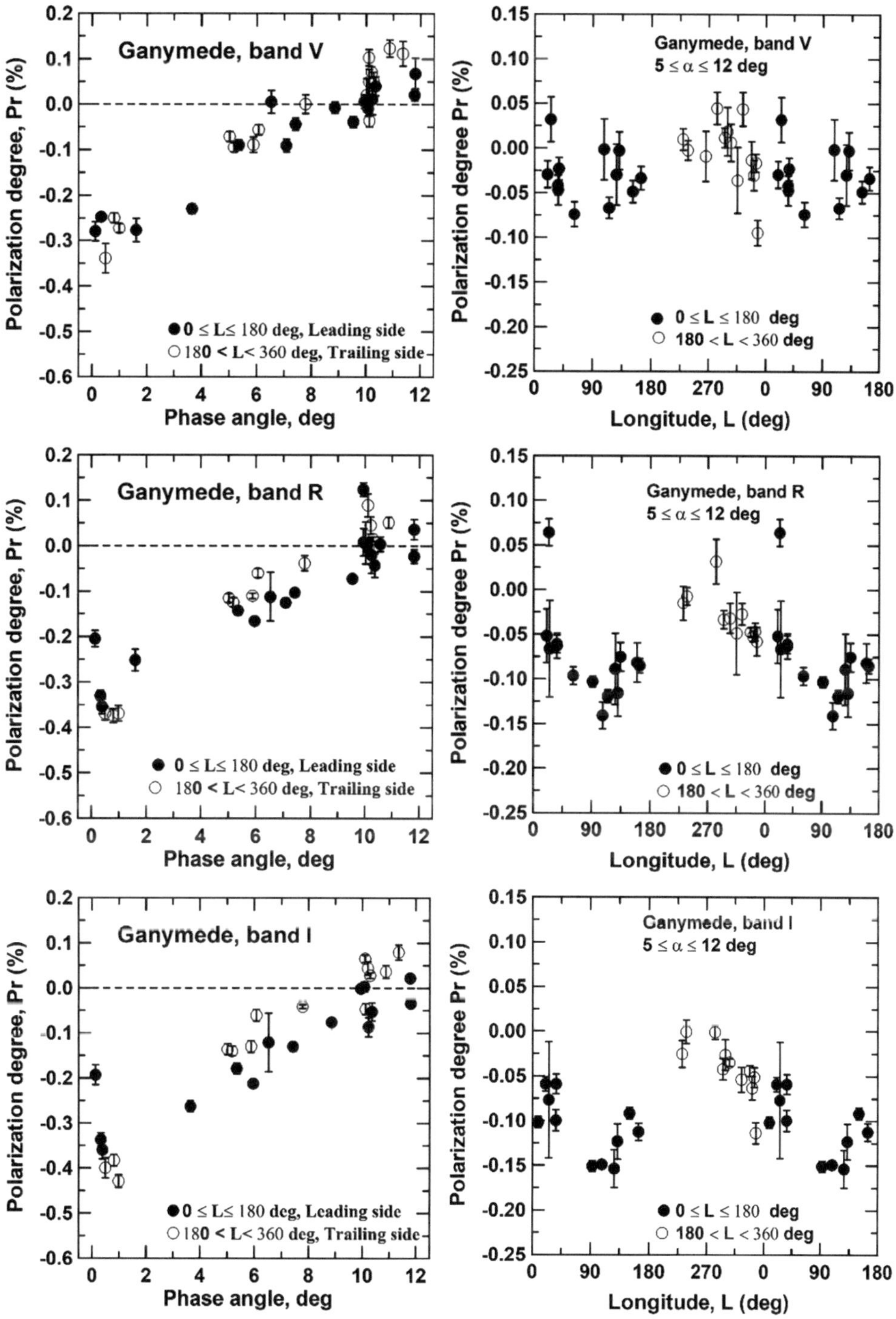

Figure 7.
Left panels: Phase dependences of Ganymede's polarization degree in the VRI bands. Filled circles – data for the leading hemispheres, 0° ≤ L ≤ 180°; open circles – data for the trailing hemispheres, 180° < L < 360°. We combine results from literature [15] with additional observations obtained in the period 2022–2023. Right panels: Longitudinal dependences of Ganymede's polarization after accounting for linear approximations to the observed PPC of each hemisphere over α = 5°–12°.

At phase angles $\alpha \approx 2°-12°$, the leading side is darker than the trailing one [7, 24, 25]. However, near opposition ($\alpha < 2°$), the brightnesses of the two hemispheres become comparable; consequently, the amplitude of the BOE for the leading hemisphere exceeds that for the trailing hemisphere. Following [23], we summarize the principal characteristics of Callisto's polarized radiation needed for comparison with the other Galilean satellites.

The leading and trailing hemispheres differ markedly in the parameters of the NPB over different phase-angle intervals (see **Figure 8**, lower left panel). For the leading, darker hemisphere, $|P_{min}|$, α_{min}, and α_{inv} are larger than for the trailing hemisphere. Starting from $\alpha \approx 3°$, an inversion occurs – the degree of polarization of the leading side becomes systematically smaller in absolute value than that of the brighter trailing side, consistent with the change in the hemispheric brightness ratio.

After correcting the observed polarization (the procedure is described in Section 3 and in the study by Rosenbush et al. [23]), the LDP at a phase angle of 6° appears as a sinusoid-like function with an amplitude of about 0.2% and a polarization minimum of $\approx -0.8\%$ at a longitude of $\approx 90°$, that is, on the leading hemisphere.

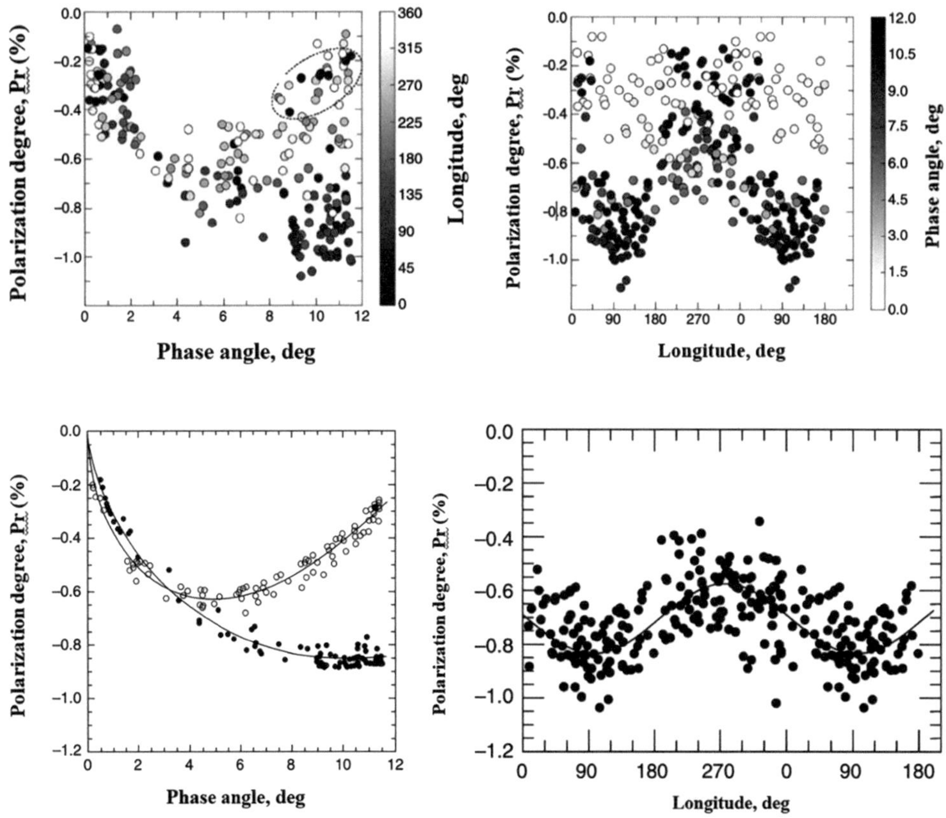

Figure 8.
Top row: The observed degree of linear polarization for Callisto (V band) as a function of phase angle (the left-hand panel) and longitude of the satellite's central meridian (the right-hand panel). Closed circles – data for the leading side (0° ≤ L ≤ 180°); open circles – data for the trailing side (180° < L < 360°). Bottom row: The left-hand panel – NPB for the leading and trailing hemispheres of Callisto after the correction for the orbital longitudinal variations. Solid curves represent the best fit to the data by a trigonometric expression of Lumme and Muinonen [26]. The right-hand panel – longitude dependence of polarization for Callisto in the V filter at the phase angle α = 6° after accounting for the phase-angle dependence of polarization.

Differences in the morphology of the NPBs of the two hemispheres and in their brightness amplitudes may be caused by differences in the microstructure and composition of their surface layers. The rather deep negative branch indicates that the surface is not only microscopically rough but also covered with fine, dark dust. High-resolution images of Callisto obtained by Galileo support this conclusion: the surface is heavily cratered, and much of the icy terrain is mantled by dark material. According to Klemaszewski et al. [27], these surface characteristics result from erosion of the upper layer due to the sublimation of volatiles.

4.2 Saturn's satellites

4.2.1 Rhea

Rhea, the second-largest satellite of Saturn, is one of the high-albedo airless bodies of the solar system ($\rho_V = 0.95$) that exhibits a BOE. Its surface is covered with water ice [28, 29]. Images of the satellite obtained by the Voyager spacecraft showed that the surface of Rhea's trailing hemisphere is geologically younger than that of the leading hemisphere. The leading hemisphere is brighter than the trailing one [30]. Therefore, one might expect the polarization degrees of the two hemispheres to differ. Polarimetric observations of Rhea are extremely scarce, owing to the challenging observing conditions caused by the satellite's proximity to Saturn's disk. Zaitsev [31] reported preliminary data on Rhea's longitudinal polarization dependence (LDP) in the V band.

According to our data (see **Figure 9**), the amplitude of Rhea's LDP in the V band does not exceed 0.02%, but it increases with wavelength and becomes significant at ≈0.1% and ≈0.15% in the R and I bands, respectively.

4.2.2 Dione

Dione ($\rho_V = 0.998$) is compositionally similar to Rhea, with heavily and weakly cratered regions. The leading side of the satellite is about 25% brighter than the trailing side, possibly due to intensive micrometeoritic bombardment on this hemisphere [32]. Kulyk [33] measured the polarization of Dione and Rhea and found that the PPCs of their leading and trailing hemispheres display deep minima of negative polarization (between −1.5% and −2.0%) at phase angles below 2°. However, with uncertainties of 0.1%–0.3% and limited coverage in phase angle and longitude, the available data were insufficient to define the phase dependence with confidence. **Figure 10** shows the results of our measurements of the phase (left panel) and longitudinal (right panel) dependences of Dione's polarization in the R band. Both figures indicate the absence of a noticeable longitudinal dependence of Dione's polarization.

Despite the fact that Dione is similar to Rhea, there is no significant longitudinal dependence on the polarization.

4.2.3 Iapetus

Iapetus exhibits a pronounced albedo dichotomy: a dark leading hemisphere (geometric albedo ≈0.04) and a bright trailing hemisphere (≈0.39). On the bright side, a narrow polarization opposition effect (POE) is observed near opposition:

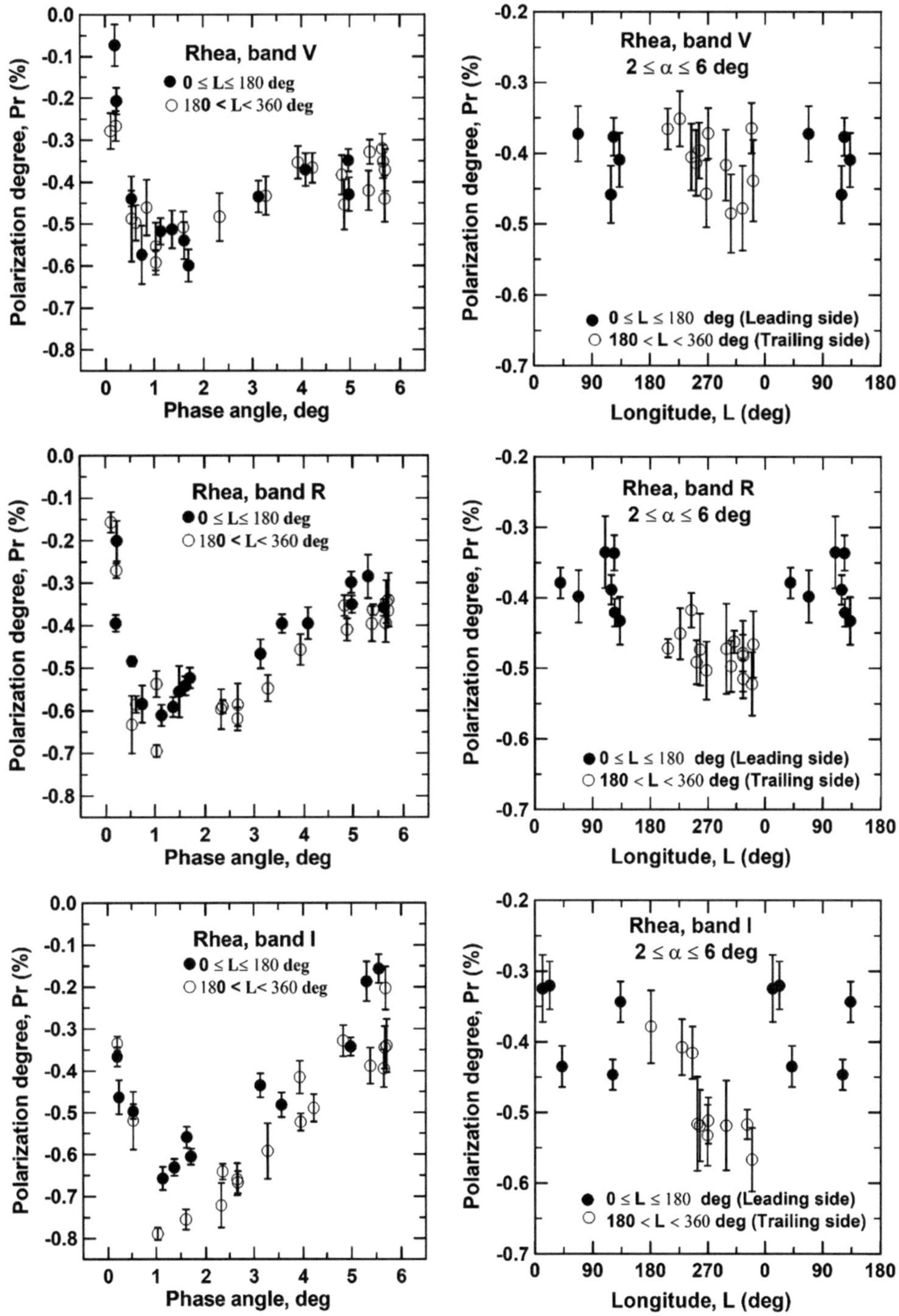

Figure 9.
Left panels: Polarization phase curves in the V (top), R (middle), and I (bottom) bands. Filled circles – data for the leading sides, 0° ≤ L ≤ 180°; open circles – data for the trailing sides, 180° < L < 360°. Right panels: Longitudinal dependences of Rhea's polarization (LDP) after applying a linear approximation to the phase curve of each hemisphere over α = 2°–6°.

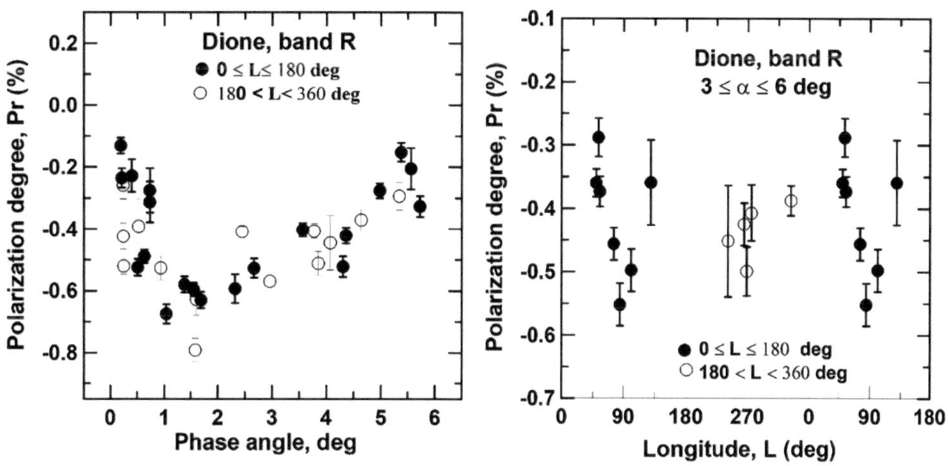

Figure 10.
Dione. Left panel – phase dependence of polarization in the R band. Filled circles – data for the leading side, 0° ≤ L ≤ 180°; open circles – data for the trailing side, 180° < L < 360°. Right panel – longitudinal dependence of Dione's polarization after accounting for a linear approximation to the phase curve of each hemisphere over α = 3°–6°.

at $\alpha = 0.77^{\circ}$ the polarization is $P \approx -0.9\%$, decreasing to $\approx -0.3\%$ by $\alpha \approx 5.2°$ (in all BVRI bands), with a very weak spectral dependence [34]. On the dark side, by contrast, as the phase angle increases from ≈0.5° to ≈6.0°, the absolute value of the negative polarization grows from ≈0.5% to ≈1.3%, with the branch minimum near $\alpha \approx 5.9°$. Over the same phase interval, 3°–6°, the dark side is polarized (in absolute value) roughly three times more strongly than the bright side; for $\alpha < 1°$, the bright side is more deeply polarized, consistent with a POE for a high-albedo surface [35].

Scattering-model inversions using the radiative-transfer coherent-backscattering approach (RT–CB) and MSTM confirm differing surface properties on the two hemispheres. RT–CB yields geometric albedos of ~0.40 for the bright (trailing) side and ~0.08 – 0.10 for the dark (leading) side, with inversion angles α_{inv}~7° (bright) and ~11°–13.5° (dark). MSTM fits indicate characteristic particle radii of ~0.10–0.20 μm; for the dark material, an effective refractive index consistent with a mixture of Fe_2O_3 + Fe + H_2O, whereas weakly absorbing ice suffices for the bright side. In the models, stronger absorption and/or reduced multiple scattering make the negative branch shallower, explaining the contrasting polarization behavior of the two hemispheres [36].

A strong LDP on Iapetus has long been noted, but on the leading side, it is difficult to separate from the phase trend because of geometric constraints (ideally, one needs a series at fixed phase and varying longitudes). This motivates continued polarimetric monitoring.

Figure 11 shows the phase and longitudinal dependences of Iapetus's polarization in the V and R bands, obtained at CRAO and Peak Terskol Observatory and compiled from the literature [31, 35–37]. A linear approximation to the phase curve was fitted over α = 3°–6°; after removing this trend, the LDP is evaluated at a fixed phase angle of 4°.

After linear detrending of the phase curves over $\alpha = 3°–6°$, the LDP is referenced to $\alpha = 4°$. As seen in the figure, the degree of polarization of the leading, dark side at longitudes $L \approx 90°$ is about −1.3%, while for the trailing, bright side at $L \approx 270°$ it is

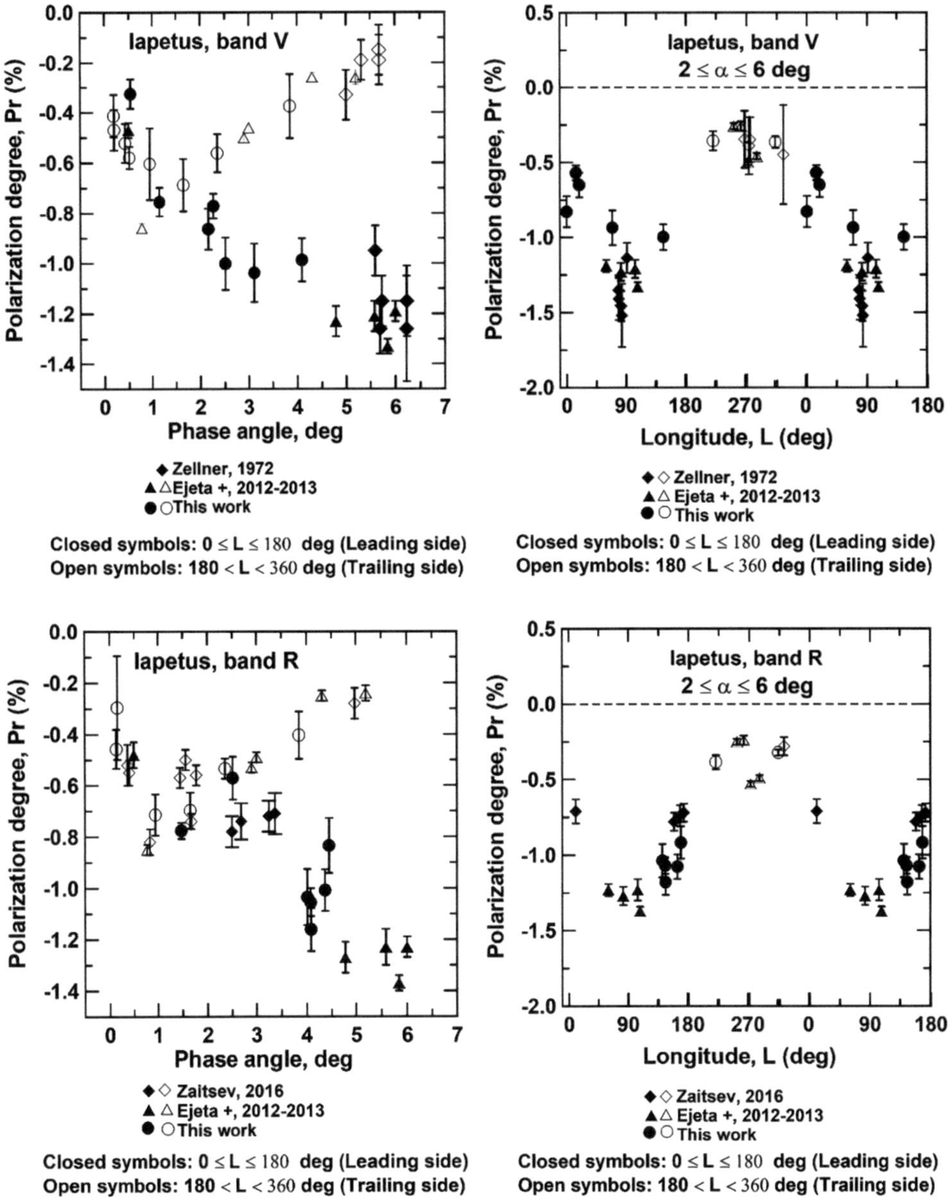

Figure 11.
Phase (left panels) and longitude (right panels) dependencies of Iapetus's polarization in the V and R bands.

about −0.3%. The large LDP amplitude reflects the strong albedo contrast between the two hemispheres, as illustrated in **Figure 12**.

5. Discussion

The principal photometric and polarimetric characteristics of the satellites' hemispheres are summarized in **Table 1**.

Figure 12.
Map of Iapetus showing the distribution of dark (leading side) and bright terrains (trailing side) across the surface [37].

Object	Region of α	Albedo's dichotomy of surface	Morphology of LDP
Jupiter family			
Europa	Positive branch: $5 \leq \alpha \leq 12°$ $<\alpha> = 9°$	The leading side is brighter than the trailing side	Polarization degree of the leading side is less than that of the trailing side
Io	Negative branch: $10 \leq \alpha \leq 12°$	Unknown	The LDP is insignificant
Ganymede	Negative branch: $5 \leq \alpha \leq 12°$ $<\alpha> = 8°$	The leading side is brighter than the trailing side	The $\|P_r\|$ of the leading side is larger than that of the trailing side
Callisto	Negative branch: $<\alpha> = 6°$	The leading side is darker than the trailing side	The $\|P_r\|$ of the leading side is larger than that of the trailing side
	Negative branch: $\alpha \leq 3°$	The leading side is brighter than the trailing side	The longitudinal dependence is not determined due to the lack of a sufficient number of observations
Saturn family			
Rhea	Negative branch: $2 \leq \alpha \leq 6°$ $<\alpha> = 4°$	The leading side is brighter than the trailing side	The $\|P_r\|$ of the leading side is smaller than that for the trailing side
Dione	Negative branch: $2 \leq \alpha \leq 6°$ $<\alpha> = 4°$	The leading side is brighter than the trailing side	There is no longitudinal dependence
Iapetus	Negative branch: $2 \leq \alpha \leq 6°$ $<\alpha> = 4°$	The leading side is darker than the trailing side	The $\|P_r\|$ of the leading side is larger than that of the trailing side

Table 1.
Albedo's dichotomy and morphology of LDP, observed for the leading and trailing sides of the selected Jupiter's and Saturn's satellites in the opposition region.

Composition (ratio of volatile to non-volatile components), surface structure (degree of cratering), magnetospheric processing, and the leading–trailing asymmetry of meteoritic erosion are the main factors shaping differences in regolith particle properties – and hence the observed photometric and polarimetric behavior. Because these factors may act in combination, disentangling them solely from observations is often difficult. Nonetheless, several patterns can be identified using the data in **Table 1**.

For Europa, the photometric and longitudinal polarization changes on the positive branch of the phase curve are consistent with Umov's law [38, 39]. According to Buratti [25], for Europa, which was previously shown to have a less compacted trailing side, magnetospheric alterations predominate.

Io, by contrast, shows a small LDP: over $\alpha \approx 4°–12°$ its amplitude did not exceed 0.1% during our observing interval, yet it was notably larger about 50 years ago. For a continually renewed volcanic surface with a sublimation–volcanic atmosphere, local and temporal changes in surface-layer composition can lead to both short-term and long-term variations in LDP.

Ganymede exhibits the most prominent LDP among the icy Galileans. The amplitude increases with wavelength and reaches ~0.15% in the I band, while a net longitudinal modulation of ~0.2% is robustly retrieved after removing $P(\alpha)$. The negative branch $P(\alpha)$ has a single minimum at small phase angles; together with independent constraints on the ice/dark-contaminant distribution, this indicates genuine longitudinal heterogeneity of the regolith.

Callisto's phase curve does not display a POE at exact backscattering, but it shows substantial differences in the negative-branch parameters between the leading and trailing sides. Buratti [25] attributes the difference in Callisto to enhanced meteoritic erosion on the leading side. The amplitude of the orbital modulation grows with α. These traits are consistent with differing microtexture/albedo of the two hemispheres' regolith and point to marked longitudinal optical heterogeneity.

For Saturn's satellite Rhea, despite the challenging proximity to Saturn's disk, earlier work noted substantial absolute minima of $P(\alpha)$ at small phase angles and preliminary indications of longitudinal variations – underscoring the need for systematic longitude coverage at fixed α.

Unlike Rhea, Dione has not been found to exhibit an LDP, at least in the R band. At the same time, both satellites have leading hemispheres that are lighter than their trailing ones.

For Iapetus, a narrow POE is found on the bright (high-albedo) side at small α, whereas on the dark side, the negative branch is deep and strengthens with phase; this demonstrates that the LDP here directly reflects the albedo dichotomy and the associated differences in regolith microphysics.

6. Conclusion

The analysis of high-precision polarimetric observations from 2018 to 2025, obtained with the two-channel POLSHAKH polarimeters, demonstrates convincingly that a number of airless satellites of Jupiter and Saturn exhibit longitudinal variability of surface polarization (LDP) superposed on the phase dependence $P(\alpha)$. The adopted decomposition of $P(\alpha, L)$ into a phase term and an orbital sinusoidal modulation, followed by linearization over a narrow α-range, enables the extraction of LDP amplitudes at the level of a few hundredths of a percent and allows a like-for-like comparison of the leading and trailing hemispheres at similar phase angles.

Overall, the results support the following:

(1) The LDP is an informative diagnostic of optical and structural inhomogeneities in the regolith of airless bodies.

(2) The amplitude and spectral behavior of the LDP differ substantially among satellites and correlate with geological and geophysical factors (icy versus dark contaminants, volcanism, albedo dichotomies, magnetospheric processing, and meteoritic erosion).

(3) Robust reconstruction of the polarization phase dependence (PDP) and the LDP from the observed $P(\alpha, L)$ requires multiband time series spanning wide ranges of phase angles and longitudes – especially critical for systems with temporally variable surfaces (e.g., Io) and for objects with a limited observable phase domain (Saturnian satellites).

These conclusions set the stage for further joint analyses of photometry, spectroscopy, and polarimetry, combined with physically consistent scattering models (RT–CB), to retrieve regolith microstructural parameters and to map disk-resolved optical inhomogeneities.

Author details

Nikolai Kiselev[1,2*] and Marina Shcherbina[2]

1 Crimean Astrophysical Observatory, Nauchny, Russia

2 Institute of Astronomy of the Russian Academy of Sciences, Moscow, Russia

*Address all correspondence to: kiselevnn42@gmail.com

References

[1] Clarke D. Stellar Polarimetry. Weinheim: Wiley-VCH; 2010. p. 415

[2] Shakhovskoy DN, Kiselev NN, Dolgopolov AV, Antonyuk KA, YuS I, Karpov NV, Taradiy VK, Savushkin AA, Ryabov AV, Taran AV. Dual-channel photoelectric polarimeters (POLSHAKH) of the Crimean Astrophysical Observatory and the terskol peak observatory: Basic design, construction, and first observation results. INASAN Science Reports. 2024;**9**(4):165–176. DOI: 10.51194/INASAN.2024.9.4.009

[3] JPL Solar System Dynamics Group. Horizons System: Web Interface [Internet]. 2025. Available from: https://ssd.jpl.nasa.gov/horizons/ [Accessed: 2025-September-01]

[4] Cassen P, Reynolds RT, Peale SJ. Is there liquid water on Europa? Geophysical Research Letters. 1979;**6**:731. DOI: 10.1029/GL006i009p00731

[5] Sparks WB, Hand KP, McGrath MA, et al. Probing for evidence of plumes on europa with HST/STIS. The Astrophysical Journal. 2016;**829**:121

[6] Sparks WB, Schmidt BE, McGrath MA, et al. Active cryovolcanism on Europa? The Astrophysical Journal Letters. 2017;**839**:L18

[7] Thompson DT, Lockwood GW. Photoelectric photometry of Europa and Callisto 1976–1991. Journal of Geophysical Research: Planets. 1992;**97**:14761–14772

[8] Dollfus A. Optical polarimetry of the Galilean satellites of Jupiter. Icarus. 1975;**25**:416–431. DOI: 10.1016/0019-1035(75)90006-8

[9] Chigladze RA. Investigation of the Polarimetric Properties of the Galilean Satellites of Jupiter and the Planet Uranus [thesis]. Abastumani: Abastumani Astrophysical Observatory; 1989. p. 175

[10] Kiselev N, Rosenbush V, Muinonen K, Kolokolova L, Savushkin A, Karpov N. New polarimetric data for the Galilean satellites: Europa observations and modeling. The Planetary Science Journal. 2022;**3**(6):134. DOI: 10.3847/PSJ/ac6bef

[11] Williams DA, et al. Global geologic mapping of Io: Preliminary results. In: Abstracts of the Annual Meeting of Planetary Geologic Mappers; 2008; Flagstaff (AZ)

[12] Lopes-Gautier R, McEwen AS, Smythe WB, et al. Active volcanism on Io: Global distribution and variations in activity. Icarus. 1999;**140**:243–264

[13] Geissler P, McEwen A, Phillips C, Keszthelyi L, Spencer J. Surface changes on Io during the Galileo mission. Icarus. 2004;**169**:129–164. DOI: 10.1016/j.icarus.2003.09.024

[14] Smith BA, Soderblom L, Beebe R, et al. Encounter with Saturn: Voyager 1 imaging science results. Science. 1981;**212**:163–191

[15] Kiselev N, Rosenbush V, Leppälä A, Muinonen K, Kolokolova L, Savushkin A, Karpov N. New polarimetric data for the Galilean satellites: Io and Ganymede

observations and modeling. The Planetary Science Journal. 2024;**5**:10. DOI: 10.3847/PSJ/ad0bf9

[16] Kiselev NN, Dyachenko HG, Karpov NV, Antoniuk KA. Changes in the longitude polarization dependence of Jupiter's moon Io as evidence of the long-term variability of its volcanic activity. Icarus. 2025;**426**:116351. DOI: 10.1016/j.icarus.2024.116351

[17] Veverka J, Burns JA, editor. Planetary Satellites (IAU Colloquium 28). Tucson: University of Arizona Press; 1977. p. 210

[18] Zellner B, Gradie J. Polarimetric observations of Io at 0.52 μm. In: Burns JA, editor. Planetary Satellites (IAU Colloquium 28). Tucson: University of Arizona Press; 1977. p. 407–418

[19] Veverka J. Polarization measurements of the Galilean satellites of Jupiter. Icarus. 1971;**14**:355–359. DOI: 10.1016/0019-1035(71)90006-6

[20] Mekler Y, Eviatar A. Time analysis of volcanic activity on Io by means of plasma observations. Journal of Geophysical Research: Space Physics. 1980;**85**:1307–1310. DOI. 10.1029/JA085iA03p01307

[21] G. SR, et al. Volcanic eruptions on Io. Journal of Geophysical Research: Space Physics. 1981;**86**:8593–8620

[22] USGS Astrogeology Science Center. Products [Internet]. Available from: https://astrogeology.usgs.gov/search/map/io_voyager_galileo_ssi_false_color_global_mosaic_1km [Accessed: 2025-September-14]

[23] Rosenbush VK, Kiselev NN, Afanasiev VA. Icy moons of the outer planets. In: Kolokolova L, Hough J, A-Ch L-R, editors. Polarimetry of Stars and Planetary Systems. Cambridge: Cambridge University Press; 2015. p. 340–359

[24] VI S, Vidmachenko A. Spectrophotometry and surface geology of the Galilean Jupiter's satellite Europa. In: Near-Earth Astronomy – 2015: Proceedings of the International Conference; 31 August–5 September 2015; Moscow. Russia: Terskol, Yanus-K; 2015. p. 146–149

[25] Buratti BJ. Ganymede and Callisto: Surface textural dichotomies and photometric analysis. Icarus. 1991;**92**:312–323. DOI: 10.1016/0019-1035(91)90054-W

[26] Lumme K, Muinonen K. A two-parameter system for linear polarization of some solar system objects. In: Harris AW, Bowell E, editors. Abstracts for IAU Symposium 160: Asteroids, Comets, Meteors 1993. Belgirate, Italy. Houston (TX): Lunar and Planetary Institute (LPI Contribution No. 810); 1993. p. 194–199

[27] Moore JM, Chapman CR, Bierhaus EB, Greeley R, Chuang FC, et al. Callisto. In: Bagenal F, Dowling TE, McKinnon WB, editors. Jupiter: The Planet, Satellites and Magnetosphere. Boulder, Colorado, USA: Cambridge University Press; 2004. p. 397–426

[28] Fink U, Larson HP, TN III G, Treffers RR. Infrared spectra of the satellites of Saturn: Identification of water ice on Iapetus, Rhea, Dione, and Tethys. The Astrophysical Journal. 1976;**207**(2):L63–L67

[29] Clark RN, Steele A, Brown RH, et al. Saturn's satellites: Near-infrared spectrophotometry (0.65–2.5 μm) of the

leading and trailing sides and compositional implications. Icarus. 1984;**58**:265–281

[30] Harris DL. Photometry and colorimetry of planets and satellites. In: Kuiper GP, Middlehurst BM, editors. Planets and Satellites. Chicago: University of Chicago Press; 1961. p. 272–342

[31] Zaitsev S. The Negative Polarization of the Scattered Radiation of Selected Atmosphereless Solar System Bodies. [thesis]. Kyiv: The Main Astronomical Observatory of the National Academy of Sciences of Ukraine; 2016

[32] Buratti BJ. Planetary satellites. In: Lucy-Ann M, P.r. W, T.v J, editors. Encyclopedia of the Solar System. 2nd ed. Boston: Academic Press; 2007. p. 365–382

[33] Kulyk I. Brightness and polarization opposition effects at low phase angles of the Saturnian satellites Tethys, Dione, and Rhea. Planetary and Space Science. 2012;**73**:407–424. DOI: 10.1016/j.pss.2012.07.019

[34] Ejeta C, Boehnhardt H, Bagnulo S, Tozzi GP. Spectro-polarimetry of the bright side of Saturn's moon Iapetus. Astronomy & Astrophysics. 2012;**537**: A23. DOI: 10.1051/0004-6361/201117870

[35] Ejeta C, Boehnhardt H, Bagnulo S, Muinonen K, Kolokolova L, Tozzi GP. Polarization of Saturn's moon Iapetus. II. Comparison of the dark and the bright sides. Astronomy & Astrophysics. 2013;**549**:A61. DOI: 10.1051/0004-6361/201220177

[36] Ejeta C, Muinonen K, Boehnhardt H, et al. Polarization of Saturn's moon Iapetus. III. Models of the bright and the dark sides. Astronomy & Astrophysics. 2013;**554**:A117

[37] USGS Astrogeology Science Center. Iapetus Voyager Airbrush Global Mosaic 783 m [Internet]. Available from: https://astrogeology.usgs.gov/search/map/iapetus_voyager_airbrush_global_mosaic_783m [Accessed: 2025-September-01]

[38] Zellner B. On the nature of Iapetus. The Astrophysical Journal. 1972;**174**: L107–L109. DOI: 10.1086/180959

[39] Umov N. Chromatische depolarisation durch Lichtzerstreuung. Physikalische Zeitschrift. 1905;**6**: 674–676

Section 3

Gas Processes

Chapter 4

Planetary Nebulae as Agents of Interstellar Enrichment and Stellar Evolution Insight

Şengül Yalgın

Abstract

Planetary nebulae (PNe) represent a crucial evolutionary phase in the life cycle of intermediate-mass stars, during which significant amounts of chemically enriched gas and dust are returned to the interstellar medium (ISM). This chapter reviews mass-loss processes in PNe through a historical lens, showing how advances in observations and theory have shaped the current understanding of stellar-envelope ejection. Using a sample of 16 well-studied PNe, we estimate dust-to-gas mass ratios, total ejected masses in solar units, and grain abundance relative to the ISM to quantify their contribution to ISM enrichment. These examples illustrate the diversity and scale of PN feedback into the Galactic environment. We also highlight the molecular and chemical significance of the returned material and its role in setting the initial conditions for future star and planet formation. The chapter contextualizes the PN phase as a meaningful contributor to the Galactic matter cycle, particularly in the early stages of stellar and planetary genesis.

Keywords: planetary nebula 1, ISM 2, dust 3, grain abundance 4, depletion 5, dust fraction

1. Introduction

Planetary nebulae (PNe) represent a short–lived phase in the life of low- and intermediate–mass stars (with initial masses of about 0.8–8 $M_{\odot}$), yet they provide key insights into both stellar evolution and the chemical enrichment of galaxies. Their diagnostic value arises not only from the ionized gas but also from the dust component, which carries essential information on the expelled material and its subsequent evolution.

At the end of the Asymptotic Giant Branch (AGB), dying stars undergo intense mass loss through slow, dense winds that expel chemically processed gas and newly formed dust into the surrounding environment [1]. Their interiors at this stage are structured around a degenerate carbon-oxygen core, enclosed by alternating helium- and hydrogen-burning shells beneath a deep convective envelope [2]. As the stellar envelope is progressively removed, the star evolves toward the PNe phase, establishing the link between late stellar evolution and the enrichment of the interstellar medium (ISM). Material released by stellar winds accumulates in circumstellar shells, where the combination of high densities and low temperatures creates ideal conditions for molecules to condense into solid dust grains in significant quantities [3].

The chemical composition of the dust depends strongly on the progenitor star's surface chemistry. Carbon-rich nebulae tend to contain graphite, amorphous carbon, or polycyclic aromatic hydrocarbons (PAHs), while oxygen-rich nebulae show various silicate forms [4]. This contrast provides a direct link between nucleosynthesis in the progenitor star and the type of solid particles released into the surrounding medium.

This chapter adopts a historical lens to show how observational and theoretical advances converged toward our present understanding of stellar-envelope ejection. Using a set of 16 well-studied PNe, we estimate dust-to-gas mass ratios and total ejected masses in solar units to assess their contribution to ISM enrichment. The sample illustrates the diversity and scale of PN feedback and provides a consistent basis for comparison with literature values.

Beyond the census of masses, we highlight the molecular and chemical significance of the returned material. The injected dust and gas alter the composition of the surrounding medium and help set the initial conditions for subsequent star and planet formation. Throughout the chapter, we emphasize the continuity between post-AGB dust formation and the PN phase, tracing how evolved-star outflows shape the composition of material later returned to the ISM.

This study focuses on a homogeneous set of 16 compact PN, whose parameters have been consistently derived within the same modeling framework. The sample provides a controlled reference for testing dust–gas correlations and benchmarking depletion behavior, while integrating extensive contextual synthesis. The discussion of expected molecular species was developed to infer plausible dust and gas compositions based on elemental abundances, even in the absence of direct infrared spectral data.

This chapter begins with a concise historical overview of planetary nebula research. Section 3 examines the main spectral features that trace dust and molecular components. Section 4 outlines the path of gas and dust injection into the ISM. Section 5 presents quantitative results from the 16-object sample, derived through photoionization modeling. Section 6 discusses the broader implications for dust evolution and ISM enrichment, and Section 7 summarizes the main findings.

2. Timeline of planetary nebula research

From discovery to diagnosis: two centuries of understanding Planetary Nebulae

- **1781 [Observation]** – [5] published a nebula catalogue including M27, M57, M76, and M92. William Herschel classified them as "planetary nebulae" due to their round, planet-like shapes.

- **1864 [Observation]** – [6] obtained the first spectrum of NGC 6543. The detection of bright emission lines showed that planetary nebulae consist of gas, not unresolved stars.

- **1922 [Observation]** – [7] demonstrated a correlation between central star brightness and nebular size, suggesting that the energy source is the central star.

- **1926 [Theory]** – [8] showed that ionization in planetary nebulae occurs through Lyman continuum photons.

- **1928 [Method]** – [9] introduced a method to estimate stellar temperatures using the Hβ line, later known as the "Zanstra method".

- **1928 [Theory]** – [10] identified the so-called "nebulium" lines as forbidden transitions of ions such as [N II] and [O III], thereby correcting decades of misinterpretation.

- **1956 [Theory]** – [11] proposed that planetary nebulae represent a short-lived evolutionary phase of red giants evolving into white dwarfs. This scenario highlighted the evolutionary place of planetary nebulae but also revealed inconsistencies between observed expansion ages and theoretical heating timescales.

- **1960s–1970s [Observation]** – Infrared and millimeter observations revealed that AGB stars underwent intense mass loss. The presence of circumstellar dust shells confirmed that planetary nebula progenitors lost a significant fraction of their mass before nebula formation.

- **1966 [Theory]** – [1] showed that planetary nebulae are formed through the ejection of red giant envelopes and suggested external mechanisms, such as instabilities and radiation pressure, to explain mass loss.

- **1971 [Theory]** – [12] introduced a model of post-AGB evolution with double-shell burning. He showed that PN nuclei maintained nearly constant luminosity while heating up and moving leftward on the H–R diagram. This resolved earlier contradictions but also predicted a critical envelope mass ($\sim 0.001 M_{\odot}$) required for PN visibility.

- **1978 [Method]** – [13] emphasized the use of spectroscopic diagnostics to determine electron temperatures and densities in nebulae.

- **1982s [Theory]** – The "Interacting Stellar Winds" model explained bright PN shells as the result of fast post-AGB winds colliding with previously ejected slow AGB winds [14].

- **1983 [Code]** – [15] released the first version of the CLOUDY code, enabling photoionization modeling of nebulae.

- **2000–2010 [Observation + Theory]** – The advent of CCD imaging and IR satellites such as IRAS, ISO, Spitzer, and WISE expanded PN studies beyond ionized shells. Observations revealed halos, molecular remnants, and dust excesses, showing that PN structures are multi-layered.

  Photoionization models (e.g., [16–19]) incorporated dust and shocks to better match observations.

- **2003–2004 [Observation]** – [20] noted the complete removal of stellar envelopes at the end of the AGB [21] and demonstrated that central star temperatures could rise to $\sim 250000\,K$during PN evolution.
- **2009 [Review]** – [22] published a major review on PN shaping, testing the binary hypothesis for morphology and evolution.
- **2010–2012 [Method]** – [23, 24] refined PN classification with spectroscopy, eliminating PN mimics [25], and developed the NEAT code, allowing automated abundance analysis from spectra.
- **2015 [Theory]** – [26] emphasized that nearly 90% of low- and intermediate-mass stars pass through the PN phase, making it a nearly universal endpoint of stellar evolution.
- **2016 [Review]** – [27] summarized advances in determining PN gaseous abundances and the challenges of abundance analysis.
- **2016–2019 [Theory]** – [28, 29] revised post-AGB evolutionary timescales, showing that central stars evolved faster than older models predicted, resolving long-standing discrepancies between PN ages and stellar heating.
- **2016–2022 [Database]** – [30–34] expanded the HASH database, compiling nearly 3000 Galactic PNe with uniform classifications.
- **2017 [Review]** – [35] published a detailed guide on nebular spectroscopy for H II regions and PNe, including the abundance discrepancy factor.
- **2022 [Review]** – [36] published a modern synthesis on PN surveys, classification, and HASH methodology (Parker, 2022).
- **2022 [Method]** – [37] compared PN abundances to Galactic chemical distributions, highlighting their role as metallicity tracers.
- **2023 [Code]** – [38] published the latest CLOUDY release, improving modeling of dust, stellar parameters, and nebular conditions.
- **2024 [Review]** – [39] presented a comprehensive historical and evolutionary review of planetary nebulae, integrating observational, theoretical, and modeling perspectives.

3. Spectral features

3.1 The 30 μm infrared bump

The broad IR emission band peaking around 30 μm, the so-called 30 μm bump, represents one of the most enigmatic dust features observed in PNe and their progenitor AGB stars. It represents thermal emission from circumstellar dust grains in radiative equilibrium at temperatures typically between 80 K and 150 K, consistent with the Wien peak of cold dust. Such emission can account for up to 40% of the nebula's total radiated energy, playing a dominant role in shaping the IR spectral energy distribution (SED) [40]. The bump originates as the ejected stellar envelope cools and expands, shifting the SED toward longer wavelengths, with the emission maximum generally falling between 25 μm and 60 μm. Observational studies using *Spitzer* and *ISO* spectroscopy have consistently detected this feature in both Galactic and Magellanic Cloud PNe [41–43].

The shape, strength, and exact peak wavelength of the 30 μm bump vary widely among different objects, reflecting differences in dust composition, temperature gradients, and evolutionary state. Carbonaceous materials such as amorphous carbon, hydrogenated amorphous carbon (HAC), silicon carbide (SiC), magnesium sulfide (MgS), and large molecular species like fullerenes have all been proposed as contributors to the mid- and far-infrared opacity that shapes this continuum. Variations in the relative abundances and optical properties of these species produce the diversity of 30 μm peaks observed across PNe. As the central star evolves and its radiation field hardens, dust grains undergo structural and chemical processing; graphitization, dehydrogenation, and sulfide condensation, that modify their emissivity and shift the IR peak. Consequently, the 30 μm bump serves as a sensitive tracer of dust temperature, composition, and processing in the late stages of stellar evolution. The principal dust components identified in planetary nebulae and their characteristic infrared bands are summarized in **Table 1**.

3.2 Magnesium sulfide (MgS)

In several carbon-rich PNe, the broad IR emission feature centered near 30 μm is attributed to MgS grains. This feature, first recognized in *IRAS* data and later confirmed by *ISO* and *Spitzer* spectroscopy, arises from lattice vibrations within MgS dust, likely condensed in the outer circumstellar envelopes of AGB stars [44, 45]. Its strength correlates with carbon-rich chemistries, suggesting that MgS forms when the C/O ratio exceeds unity and sulfur becomes available for solid-phase reactions. Observations of PNe such as NGC 7027 and IC 418 reveal that MgS often coexists with amorphous carbon and PAHs, implying a composite grain structure or sequential condensation in cooling outflows. The 30 μm band profile varies across objects, indicating differences in grain shape, coating, or temperature, typically in the range of 100–300 K.

Dust component	Characteristic bands (μm)	References
Magnesium sulfide (MgS)	26–30	[44, 45]
Silicon carbide (SiC)	11.3	[46, 47]
Fullerenes (C_{60}, C_{70})	7.0, 8.5, 17.4, 18.9	[48, 49]
Hydrogenated amorphous carbon (HAC)	3.4, 6.85, 7.25	[50–52]
Amorphous and crystalline silicates	9.7, 11, 18, 23, 28, 33	[53–56]
Polycyclic aromatic hydrocarbons (PAHs)	3.3, 6.2, 7.7, 8.6, 11.2, 12.7	[57–61]

Table 1.
Principal dust components identified in planetary nebulae and their characteristic infrared bands.

3.3 Silicon carbide (SiC)

The spectral signature of SiC is one of the most distinct markers of carbon-rich dust in PNe. SiC grains produce a prominent, broad emission feature around 11.3 μm, associated with Si–C stretching modes in crystalline or amorphous structures [46, 47]. This feature is frequently detected in the spectra of C-rich PNe observed by *ISO* and *Spitzer*, often alongside PAH and amorphous carbon bands, indicating that SiC grains form early in the condensation sequence of carbonaceous dust. Laboratory studies and meteoritic analyses have confirmed that presolar SiC grains carry isotopic anomalies characteristic of AGB nucleosynthesis, linking them directly to PN progenitors. The relative weakness of the SiC band in evolved nebulae suggests that grain surface oxidation or coating by amorphous carbon may occur as the circumstellar environment becomes more ionized.

3.4 Fullerenes

The discovery of fullerenes (C_{60} and C_{70}) in PNe marked a milestone in astrochemistry, revealing that large carbonaceous molecules can survive and form under harsh UV-irradiated conditions. The first unambiguous detection of C_{60} was made in the PN Tc 1 using *Spitzer* mid-infrared spectroscopy, which showed four strong emission bands at 7.0, 8.5, 17.4, and 18.9 μm [48]. Subsequent surveys identified fullerenes in several Galactic and Magellanic Cloud PNe, indicating that their formation is favored in moderately low-excitation and carbon-rich environments [49]. The coexistence of PAHs and fullerenes in some nebulae suggests a chemical evolution pathway in which UV photoprocessing of HAC or large PAH clusters leads to C_{60} formation. Fullerenes in PNe typically emit at temperatures of 200–400 K, reflecting stochastic heating by central star photons.

3.5 Hydrogenated amorphous carbon (HAC)

HAC represents a transitional form of carbonaceous dust composed of a mixed sp^2/sp^3 bonding network with varying hydrogen content. HAC grains are thought to form in the outer, carbon-rich regions of stellar outflows, where moderate temperatures and high densities allow the accretion of hydrogenated carbon clusters [50]. The mid-IR spectra of many PNe reveal broad plateaus between 6 and 9 μm and 11 and 15 μm, attributed to the collective vibrational modes of HAC materials rather than discrete molecular bands. These features often coexist with the Aromatic Infrared Bands (AIBs) associated with PAHs, suggesting that HAC grains act as chemical precursors to aromatic molecules. UV irradiation and thermal processing by the CS progressively dehydrogenate HAC, promoting the transformation from aliphatic to aromatic structures [51, 52]. This photochemical evolution is supported by the simultaneous presence of both aliphatic (3.4 μm) and aromatic (3.3 μm) C–H stretching features in several PNe. As the stellar radiation field hardens during the PN phase, HAC destruction and conversion into smaller PAHs and even fullerenes become increasingly efficient [48, 49].

3.6 Silicates

Oxygen-rich PNe exhibit prominent IR emission bands produced by silicate grains. The broad features centered near 9.7 and 18 μm correspond to the Si–O stretching and

O–Si–O bending modes, respectively, and are diagnostic of amorphous silicates formed in the outflows of oxygen–rich AGB stars [53, 54]. In some evolved nebulae, sharp substructures at 11, 19, 23, 28, and 33 μm have revealed the presence of crystalline silicates such as forsterite (Mg_2SiO_4) and enstatite ($MgSiO_3$) [55, 56]. The coexistence of amorphous and crystalline silicates suggests that dust processing and thermal annealing occurred during the transition from the AGB to PNe phase.

3.7 Polycyclic aromatic hydrocarbons (PAHs)

PAHs are among the most ubiquitous organic compounds detected in PNe and represent a critical stage in the carbon–dust evolution of dying stars. These large planar molecules, composed of fused benzene rings, produce a distinct set of IR emission features collectively known as the Aromatic Infrared Bands (AIBs). Prominent bands occur at 3.3, 6.2, 7.7, 8.6, and 11.3 μm, arising from C–H and C–C vibrational modes excited by UV photons from the CS [57, 58]. The relative strengths and peak positions of these bands vary systematically among PNe, tracing the ionization state, size distribution, and degree of molecular processing of the PAH population. The coexistence of PAHs with other carbonaceous species, such as HAC, SiC, and fullerenes, suggests a continuous chemical pathway driven by UV photoprocessing and grain fragmentation. Recent high-resolution observations, including JWST imaging of NGC 6537, NGC 6302, and NGC 6720, have revealed spatially confined PAH emission in photodissociation regions at the interface between ionized and molecular gas [59–61]. These findings highlight PAHs as key tracers of the complex interplay between radiation, chemistry, and dust evolution in PNe.

4. Gas and dust injection into the ISM

Dust production models for evolved star winds indicate that, in most cases, the majority of dust is released during the terminal stages of the AGB phase. For carbon stars, this outcome is particularly evident, since their dust production rate rises progressively throughout the AGB as surface carbon is enriched by recurrent mixing episodes that transport freshly synthesized material to the surface [62]. A similar trend is found in low-mass stars that never reach the C-star stage; in these systems, significant dust condensation becomes possible only in the very late AGB, when the outer layers and the circumstellar envelope cool sufficiently to permit grain formation [63]. More recently, it has been suggested that stars of about $4M_\odot$ experiencing proton-capture nucleosynthesis at the base of the convective envelope (hot bottom burning) [64], also follow this behavior.

The recognition that dust production peaks near the end of the AGB phase and the phases that follow are crucial for understanding what happens after dust formation stops. Post-AGB stars and planetary nebulae are ideal targets: their infrared excess shows the remaining circumstellar dust, and nebular lines reveal the ionized gas and total mass. Taken together, these data allow us to estimate how efficiently dust was produced at the end of the AGB and the final dust yield delivered to the ISM [65–67].

[68] report a marked augmentation in the nebular dust content with stellar mass: the dust-to-gas ratio climbs from about 5×10^{-4} at $\approx 0.9M_\odot$ to $\approx 6 \times 10^{-3}$ by $2M_\odot$.

4.1 Volatile and refractory elements

The composition of interstellar dust does not directly mirror the overall abundances in the ISM. Elements with high condensation temperatures, known as refractory species, such as Al, Ca, Fe, Mg, Si, and Ni, are efficiently incorporated into solid grains, whereas volatile elements with low condensation thresholds (e.g., C, N, O, noble gases) largely remain in the gas phase [69].

Observations reveal a clear link between the degree of elemental depletion and the condensation temperature: elements with higher condensation temperatures are strongly depleted from the gas, as they remain locked into dust grains until relatively high temperatures. Species with low condensation temperatures are easily liberated from solids under moderate heating, so their interstellar abundances are dominated by the gas phase [70].

It is worth noting that carbon and oxygen, despite being essential ingredients of dust, are formally considered volatile. Their depletions are moderate rather than extreme, but still sufficient to contribute substantially to the overall dust mass budget.

4.2 Elemental depletions

In the cold and dense regions of the ISM, refractory elements are efficiently removed from the gas phase and incorporated into solid dust grains [70]. This process, known as *elemental depletion*, hides part of the heavy-element content from direct spectroscopic detection, as these species become locked within grain lattices rather than freely circulating in the gas [71]. The degree of depletion depends strongly on the physical conditions of the medium: temperature, density, and ionization. It provides a measure of the efficiency of dust condensation [69].

In PNe, elemental depletions serve as powerful diagnostics of dust composition and formation processes. As the stellar envelope ejected during the AGB phase expands and cools, refractory elements condense into silicate, oxide, or carbonaceous solids depending on the local C/O ratio. Subsequent ultraviolet irradiation and shocks from the central star modify these grains through sputtering, oxidation, and reprocessing [72]. Comparative analyses of gas and dust abundances in PNe reveal systematic variations in depletion levels that reflect differences in morphology, excitation, and progenitor chemistry [70, 73, 74]. The characteristic properties of interstellar dust components in the THEMIS model [74] are listed in **Table 2**, offering a benchmark for interpreting the infrared features attributed to carbonaceous and silicate grains in PNe. Hence, studying elemental depletions in PNe not only constrains the mineralogical nature and survival of circumstellar dust but also links stellar ejecta to the ongoing chemical enrichment of the Galactic ISM.

4.3 PN ejecta in the ISM

Once PN ejecta leave the nebular shell, they enter the diffuse and subsequently denser phases of the interstellar medium (ISM), where gas-phase depletions provide a quantitative record of what material remains in solids. The relative fractions of major elements incorporated into solid grains are illustrated in **Figure 1**, based on

Parameter	Small C grains	Large C grains	Silicates	Total
Grain number fraction	10^9 ppb	36 ppb	80 ppb	10^9 ppb
Average radius	0.54 nm	12 nm	13 nm	0.54 nm
Average surface area	1.02 nm^2	1140 nm^2	1510 nm^2	1.03 nm^2
Average mass	1.0×10^3 amu	1.2×10^8 amu	2.5×10^8 amu	2.9×10^3 amu
Mass fraction	23%	8.1%	69%	100%
Dustiness, M_{dust}/M_{gas}	$\approx 1.3 \times 10^{-3}$	$\approx 4.4 \times 10^{-4}$	$\approx 3.8 \times 10^{-3}$	$\approx 5.5 \times 10^{-3}$
Dust-to-H mass ratio	$\approx 1.7 \times 10^{-3}$	$\approx 5.9 \times 10^{-4}$	$\approx 5.0 \times 10^{-3}$	$\approx 7.3 \times 10^{-3}$
Dust-to-H number ratio	5300 ppb	0.05 ppb	0.02 ppb	1905 ppb
Dust-to-metal mass ratio	$\approx$ 9%	$\approx$ 3%	28%	41%

Table 2.
Average physical and abundance properties of dust populations in the THEMIS model [74].

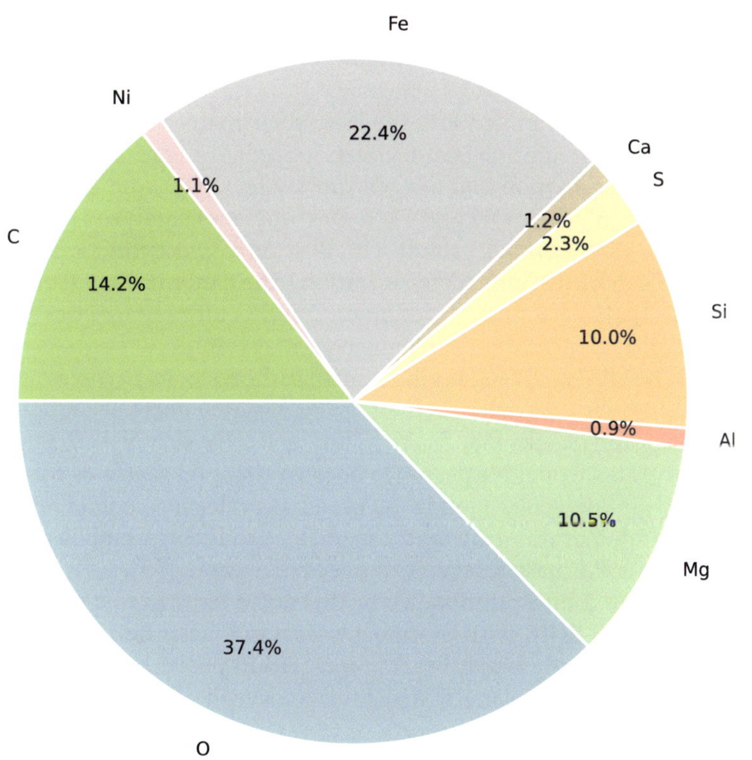

Figure 1.
Mass fraction of the various elements locked in grains. The values are adopted from [75].

the depletion patterns derived by [75]. Classical depletion analyses imply a mean dust-to-gas mass ratio near one percent, with most of the dust mass residing in silicate and oxide minerals that sequester oxygen, magnesium, silicon, and iron [76]. Carbon is more variably depleted and may populate amorphous carbon grains or PAH-like carriers [58].

In this framework, PN-injected dust does not remain static: in the diffuse ISM, it can be partially eroded by shocks and sputtering [77], whereas in denser regions, it can accrete mantles, coagulate into aggregates, and exchange material with the gas through adsorption and desorption [78]. As a result, the mineralogy and size distribution of PN grains continue to evolve toward the equilibrium set by ISM depletion patterns. When PN results are compared to this benchmark, high mass-based dust-to-gas ratios are expected to relax toward the canonical ISM value as grains mix, process, and regrow, while low ratios may instead indicate efficient dust destruction or an underestimation of the gas content.

From a broader astrophysical perspective, the key point is inheritance. If PN dust survives and becomes incorporated into cold molecular clouds, it can contribute refractory material and surface area that catalyze molecule formation and ice chemistry. Such presolar grains, now observed in meteoritic inclusions, demonstrate that some refractory AGB/PN material can indeed survive ISM processing and become incorporated into the next generation of solids [79]. The resulting mixture defines the initial solid inventory of protoplanetary disks and minor bodies, effectively linking late stellar evolution to the earliest stages of planetary assembly.

4.4 From PN dust to planetary building blocks

Once the PN ejecta are mixed into the ISM, their material begins a new cycle of chemical and physical evolution that ultimately seeds the formation of stars and planets. In diffuse gas, carbon and oxygen rapidly form CO where shielding allows, while residual oxygen enters OH and H_2O. Dust surfaces catalyze H_2 formation and serve as sites for subsequent ice growth [76, 80]. As visual extinction increases ($A_V \gtrsim 1$), freeze-out onto grain surfaces initiates the canonical ice sequence—H_2O, CO, CO_2, and CH_2OH—locking a fraction of C and O into icy mantles that later participate in dense-cloud and disk chemistry [58, 81, 82].

In low-density, UV-irradiated regions, small carbonaceous particles and PAHs are efficiently photo-processed, while refractory Fe/Mg-rich cores are more resilient against sputtering and shocks [58, 77, 78]. These selective survival effects ensure that the most robust grains from PNe persist to become the solid scaffolds for later ice and organic accretion in molecular clouds. As material cycles into colder and denser phases, these mantled grains grow and coagulate, producing composite aggregates that dominate the solid mass reservoir of pre-stellar cores [83].

When incorporated into protoplanetary disks, the same grains undergo coagulation and radial drift, with thermochemical gradients defining snowlines for volatiles [84]. Water- and carbon-bearing ices within pebbles and planetesimals deliver volatiles to forming planets, while refractory silicates and Fe-bearing inclusions set the rock–metal framework of terrestrial material [85, 86]. Thus, the refractory-rich grains and mixed carbon–oxygen solids ejected by PNe ultimately contribute to the molecular and mineralogical inventory inherited by new planetary systems, establishing a direct continuity between the final phases of stellar evolution and the earliest stages of planet formation.

4.5 Presolar grains and isotopic evidence

Laboratory analyses of presolar grains isolated from primitive meteorites demonstrate that PN dust can survive interstellar processing and become part of the

Galactic material cycle [87, 88]. These submicron- to micron-sized grains, isolated from chondritic matrices, preserve isotopic compositions that differ significantly from solar values and, therefore, record the nucleosynthetic history of their parent stars [89]. Among the most diagnostic presolar solids are SiC, graphite, and oxide grains of corundum (Al_2O_3) and spinel ($MgAl_2O_4$), along with rarer Si_3N_4, sulfide, and silicide phases. These grains display characteristic isotopic ratios, including $^{12}C/^{13}C$, $^{14}N/^{15}N$, and $^{18}O/^{16}O$, which encode the nucleosynthetic signatures of their parent AGB and PN progenitors [89–92].

In addition, many grains show enrichments in s-process nuclides such as ^{88}Sr, ^{96}Zr, and ^{138}Ba, reflecting the slow neutron-capture nucleosynthesis characteristic of late-stage AGB evolution [93, 94]. These isotopic and elemental signatures together trace the chemical evolution from AGB stars to planetary nebulae and ultimately link presolar grain populations to Galactic dust formation [92].

Micron- and submicron-sized grains originating in the winds of evolved stars and the debris of stellar explosions represent a small yet highly informative fraction of the primitive solids preserved in Solar System materials [95]. This correspondence demonstrates that dust formed in the late phases of stellar evolution can survive for long periods in the interstellar medium before its incorporation into the Solar System [96, 97]. Unlike most interstellar dust, these ancient particles largely escaped isotopic homogenization and chemical alteration during their passage through the interstellar medium and the subsequent collapse of the protosolar cloud [96].

Silicate grains dominate the recovered population, followed by refractory oxides, silicon carbide, graphite, and nitrides. In particular, presolar SiC and graphite grains contain trace-element and isotopic signatures that depend sensitively on the parent star's composition, mass-loss history, and the condensation sequence of their host minerals [96, 98]. Graphite and oxide grains reveal complementary patterns, with high $^{12}C/^{13}C$ ratios and oxygen isotopic anomalies that point to oxygen-rich or carbon-rich PN environments shaped by selective condensation and subsequent processing [87, 98]. These laboratory findings show that a portion of the PN dust inventory, particularly the most refractory silicate and carbide phases, can persist through interstellar processing, mix into molecular clouds, and ultimately seed the next generation of planetary material.

In addition to these AGB and PN sources, isotopic signatures in certain presolar grains also reveal a contribution from supernova ejecta, further underscoring the mixed origins of interstellar dust [87, 91, 99]. The α-elements (O, Mg, Si, S, Ca) are synthesized primarily in massive stars, whereas Fe- and Ni-group elements originate from Type Ia supernovae. Some SiC and graphite grain types have been traced to core-collapse supernovae; their heavy isotopic enrichments (e.g., ^{28}Si and ^{44}Ti-derived ^{41}Ca excesses [87]) attest to the diverse stellar sources contributing to the presolar dust population. Presolar grain isotopic patterns thus record inputs from both channels, linking AGB and supernova nucleosynthesis in the Galactic dust inventory.

Recent detections of interstellar objects, including 1/'Oumuamua [100], 2I/Borisov [101], and the more recent 3I/ATLAS candidate [102], provide a dynamic, modern extension to this picture. These bodies, originating beyond the Solar System, demonstrate that refractory and volatile solids formed around other stars can traverse interstellar space and survive long enough to enter new planetary systems. The peculiar, highly elongated 1I/'Oumuamua appears to represent a dense refractory remnant, possibly originating from the tidal disruption of a planetary body around a white dwarf [103] or, alternatively, from the debris of an evolved-star environment [104, 105], whereas 2I/Borisov shows a chemically distinctive coma, depleted in carbon-bearing species but enriched in highly NH_2-rich, with modest variations in C_2 and CN production [106], indicating a volatile inventory unlike that of Solar System comets, likely shaped in a distant, cold protoplanetary disk before ejection into the ISM [106–108]. Preliminary data for 3I/ATLAS indicate comparable hyperbolic motion [109] and compositional diversity [110, 111] suggesting that such interstellar travelers encompass a spectrum from rocky to icy compositions. Together, the isotopic fingerprints of presolar grains and the trajectories of these interstellar visitors highlight a continuous Galactic exchange of solid matter, from AGB stars and supernovae through the ISM to emerging planetary systems.

5. Quantitative constraints from 16 PN samples

5.1 Sample and data products

The analyses presented in this work are based on the datasets published by [112] and [113], which include optical diagnostics, distances, angular sizes, broadband infrared photometry (*IRAS* and *WISE*), and radio continuum measurements. For each planetary nebula, an individual photoionization model was recomputed using the CLOUDY code (version 23.01; 2023 release with CHIANTI atomic database updates), adopting the physical output parameters and observational constraints reported in those studies. Each simulation used the prescribed stellar and nebular inputs to recalculate the total gas mass (M_{tot}), dust-to-gas mass ratio ($\delta = M_{dust}/M_{gas}$), and relative grain compositions. The total mass of the nebula corresponds to the mass enclosed within the photoionized region and thus effectively represents the ionized gas component rather than the full nebular mass. From these primary quantities, we further calculated the dust mass fraction ($f_{dust} = M_{dust}/M_{tot}$) and derived both gas-phase and grain depletions for volatile (He, N, O, Ne, Ar, S, Cl) and refractory species (Fe, Mg, Si). Grain depletions were also calculated for O and C, which, despite their lower condensation tendencies, significantly contribute to the overall dust composition. These depletion patterns provide quantitative constraints on the efficiency of dust formation and retention within the nebular environment, offering a physical link between the ionized gas and solid-state components.

5.2 Method

Mid-infrared RGB images of the sample PNe were obtained from the HASH database, constructed using *WISE* survey data in the W4, W3, and W2 bands. These composites provide a morphological reference for the modeled sample (see **Figure 2**). In the models of [112] and [113], the nebular regions were constrained solely by ionized-gas diagnostics, as molecular and photodissociation-region (PDR) data were not available. Although mid-infrared spectroscopy is lacking, the broadband infrared and radio continuum data summarized in **Table 3** and **Table 4** serve as key constraints to calibrate the dust-to-gas ratio in the absence of full spectroscopic coverage.

In our recomputed CLOUDY models, we determined M_{tot} and δ. Using these well-defined ionized zones as a reference, we further calculated the fractional dust budget

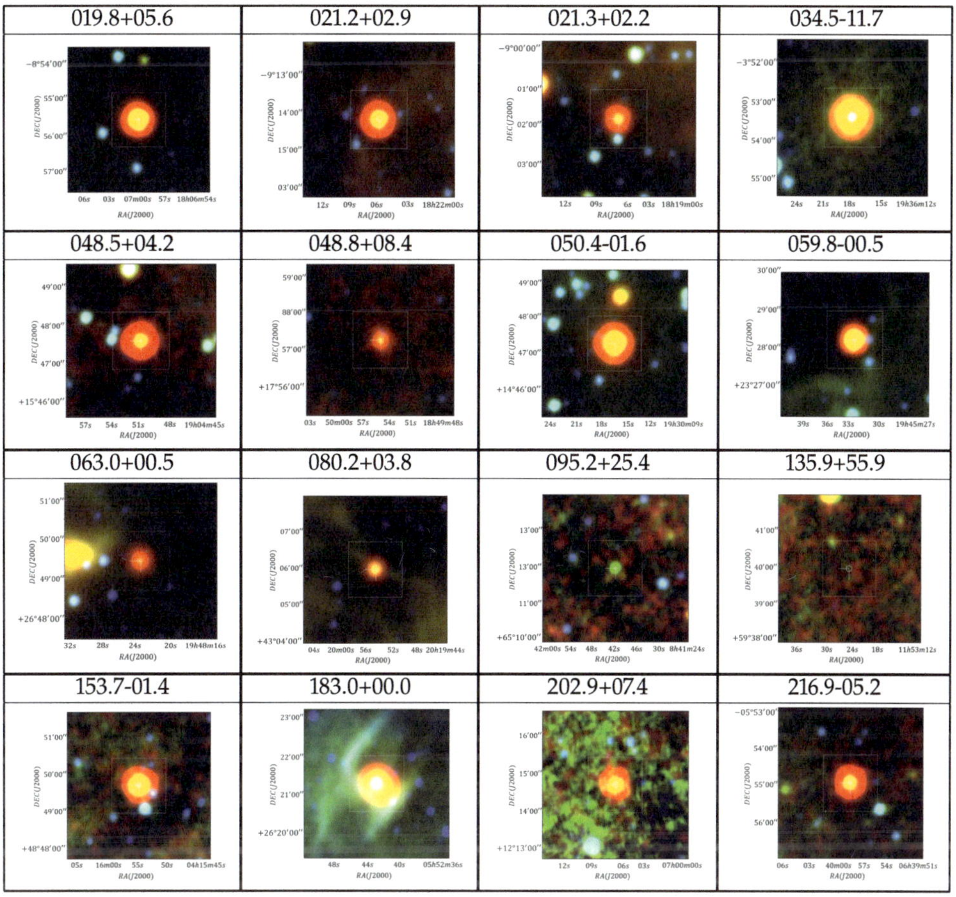

Figure 2.
RGB composite images (red = W4, green = W3, blue = W2) of 16 selected PNe were constructed using data from the WISE survey in the HASH database. Each PN appears near the center of its respective field, typically exhibiting a reddish hue. The image frames cover $500' \times 500'$ *and are aligned with north at the top and east to the left.*

and the relative elemental contributions to the grain mass, providing a theoretical framework for the expected dust composition under the prevailing ionization conditions. Finally, these mass-based parameters were interpreted in the broader context of ISM elemental depletions.

For each object, the total dust mass $M_{\rm dust}$ (derived from SED fitting) was combined with the grain-phase elemental number abundances per hydrogen atom, obtained from the CLOUDY output block *"Grain Chemical Composition"*. Let the logarithmic grain number abundance be defined as $\ell_X \equiv log_{10}[(X/H)_{\rm grain}]$. These values were converted into number abundances, weighted by the atomic masses A_X, and renormalized to obtain elemental mass fractions as follows:

$$n_X^{(\rm grain)} = 10^{\ell_X}, \tag{1}$$

$$m_X^{(\rm per\ H)} = n_X^{(\rm grain)} A_X m_{\rm u}, \tag{2}$$

PN G	WISE (Jy)				IRAS (Jy)			
	3.4 μm	4.6 μm	12 μm	22 μm	12 μm	25 μm	60 μm	100 μm
019.8 + 05.6	–	–	–	–	**4.02e-1**	3.12	4.95	**5.68e + 0**
021.2 + 02.9	2.12e-3	1.13e-3	2.22e-2	1.42e-1	–	–	–	–
021.3 + 02.2	–	–	–	–	**2.63e-1**	9.61e-1	**2.581e + 0**	**3.40e + 1**
034.5 – 11.7	–	–	–	–	**3.79e-1**	8.184e + 0	5.558e + 0	2.32e + 0
048.5 + 04.2	–	–	–	–	**2.88e-1**	2.55e-1	**4.95e-1**	**5.59e + 0**
048.8 + 08.4	2.12e-4	2.23e-4	2.985e-3	3.416e-2	–	–	–	–
050.4 – 01.6	–	–	–	–	**3.71e-1**	3.84e + 0	**2.52e + 0**	**8.59e + 1**
059.8 – 00.5	3.68e-3	1.11e-2	3.96e-1	3.90e + 0	–	4.68e + 0	4.19e + 0	5.48e-1
063.0 + 00.5	1.77e-3	3.01e-3	2.48e-2	2.46e-1	–	–	–	–
080.2 + 03.8	1.81e-3	2.83e-3	4.88e-2	2.17e-1	–	–	–	–
095.2 + 25.4	–	–	–	–	–	–	–	–
135.9 + 55.9	3.07e-5	6.28e-5	4.25e-4	2.54e-3	–	–	–	–
153.7 – 01.4	–	–	–	–	**3.94e-1**	**2.87e-1**	8.08e-1	**3.19e + 0**
183.0 + 00.0	–	–	–	–	**2.50e-1**	1.561e + 0	5.154e + 0	8.004e + 0
202.9 + 07.4	2.57e-4	5.18e-4	2.61e-3	4.99e-2	–	–	–	–
216.9 – 05.2	2.59e-3	2.42e-3	8.18e-3	1.09e-1	–	–	–	–

All flux densities are in Jy. Values in ***bold*** *denote reported upper limits (typically 3σ) from the source catalogs.*

Table 3.
Infrared flux densities for the PN sample. Upper limits are shown in ***bold***.

$$f_X = \frac{m_X^{(\text{per H})}}{\sum_Y m_Y^{(\text{per H})}}, \qquad \sum_X f_X = 1, \tag{3}$$

$$M_{\text{dust},X} = f_X M_{\text{dust}}. \tag{4}$$

Here m_u is the atomic mass unit (1.6605×10^{-24} g), and A_X are standard atomic weights (e.g., C = 12.01, O = 16.00, Mg = 24.305, Si = 28.085, Fe = 55.845), $n_X^{(\text{grain})}$ is the number abundance of element X locked in grains, relative to hydrogen, $m_X^{(\text{per H})}$ is the corresponding mass of element X per hydrogen atom, f_X is the mass fraction of element X within the total dust mixture, $\sum_Y m_Y^{(\text{per H})}$ denotes the sum of masses over all grain-forming elements Y, and $M_{\text{dust},X}$ is the total dust mass attributable to element X and represented in solar units. This procedure yields element-resolved dust masses $M_{\text{dust},X}$ ($M_\odot$).

Dust grains embedded in ionized nebulae are inferred not only from direct infrared observations but also from the observed element depletions. For any element X, the depletion factor quantifies its gas-phase underabundance relative to the solar reference and is defined as:

$$[\text{X/H}] \equiv \log(\text{X/H}) - \log(\text{X/H})_\odot, \tag{5}$$

PN G	5 GHz (Jy)	4 GHz (Jy)	1.4 GHz (Jy)
	6 cm	7.4 cm	21 cm
019.8 + 05.6	3.4e-1	–	–
021.2 + 02.9	–	1.96e + 0	–
021.3 + 02.2	–	9.57e-3	–
034.5 – 11.7	6.80e-4	–	–
048.5 + 04.2	3.00e-3	–	–
048.8 + 08.4	–	–	–
050.4 – 01.6	1.90e-2	–	–
059.8 – 00.5	2.47e-2	2.65e-2	2.00e-2
063.0 + 00.5	5.23e-3	5.96e-3	–
080.2 + 03.8	–	5.02e-3	–
095.2 + 25.4	–	–	–
135.9 + 55.9	–	–	–
153.7 – 01.4	4.00e-3	–	–
183.0 + 00.0	–	1.41e-2	–
202.9 + 07.4	–	–	–
216.9 – 05.2	–	2.48e-3	–

All flux densities are in Jy. The second header row lists the wavelengths corresponding to the stated frequencies.

Table 4.
Radio flux densities for the PN sample.

The corresponding fraction of the element locked into dust can then be expressed as:

$$\left(\frac{X}{H}\right)_{dust} = \left(\frac{X}{H}\right)_{\odot}\left(1 - 10^{[X/H]}\right), \quad (6)$$

where solar abundances are commonly adopted as the cosmic benchmark owing to their precise determination for a wide range of elements [73]. If all atoms of a species remain in the gas phase, its measured abundance should match the solar value; any observed deficiency indicates that part of the element is locked into solid dust grains.

6. Discussion

The planetary nebulae analyzed in this work span a wide dynamic range in their dust and gas properties, illustrating the diversity of condensation and destruction outcomes in evolved stellar ejecta [14, 18]. The modeled spectral energy distributions for all 16 PNe are shown in **Figure 3**, where the four outlier objects (PN G 019.8 + 05.6, 095.2 + 25.4, 135.9 + 55.9, and 080.2 + 03.8) are highlighted in purple, reflecting their distinct SED morphologies and atypical dust or gas properties relative to the main sample. The three most extreme systems; PN G 019.8 + 05.6, PN G 095.2 + 25.4, and PN G 135.9 + 55.9 define the boundaries of

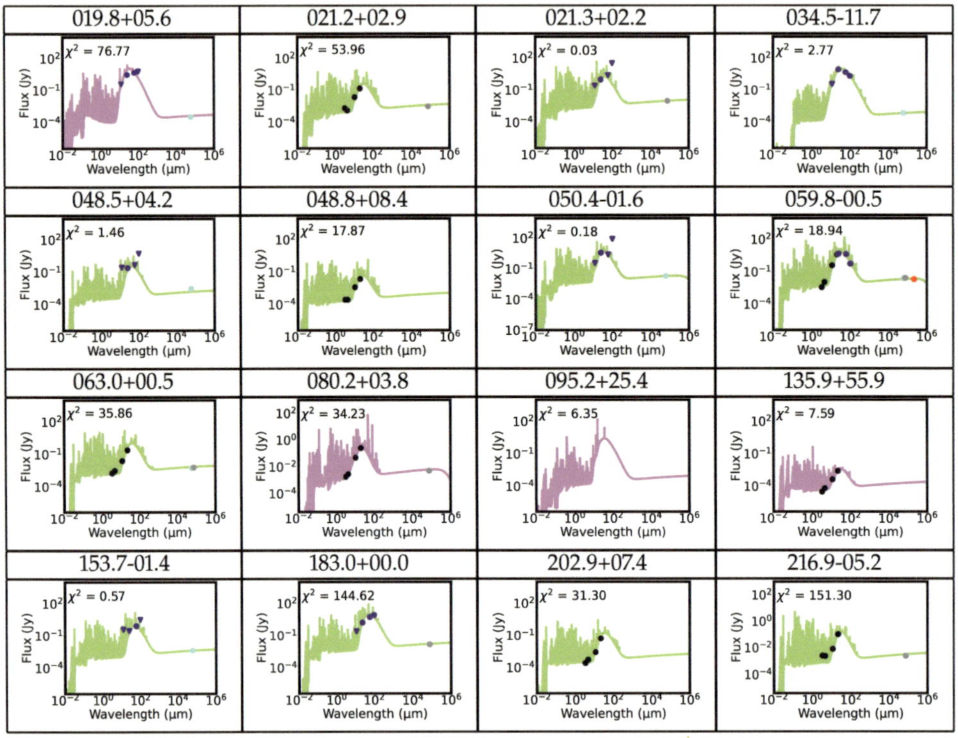

Figure 3.
Spectral energy distributions (SEDs; green and purple) generated with the cloudy models for 16 planetary nebulae (PNe). Observed IRAS, WISE, and/or radio fluxes are overplotted; triangles denote upper limits. The purple curves correspond to the outlier objects.

this behavior, from highly dust-dominated shells to genuinely dust-poor nebulae, while PN G 080.2 + 03.8 represents a chemically peculiar, oxygen-rich yet refractory-selective case. Together, these objects anchor the physical range of grain processing mechanisms encountered in our modeled sample.

It is important to emphasize that the total gas masses derived here correspond strictly to the photoionized regions modeled by CLOUDY. The computed gas masses, and hence δ should be regarded as lower and upper limits, respectively. In other words, the relatively large values of δ in **Table 4** partly reflect the restricted ionized volumes rather than intrinsically high condensation efficiencies.

Summing over all 16 objects, the ionized gas mass returned to the ISM amounts to $\langle M_{\text{gas}} \rangle = 5.5 \times 10^{-1} M_{\odot}$, and the corresponding dust yield is $\langle M_{\text{dust}} \rangle = 5.1 \times 10^{-4} M_{\odot}$. These values imply a mean dust fraction $f_{\text{dust}} \simeq 0.084$ and a mass-weighted dust-to-gas ratio $\delta_{\text{mw}} \simeq 9 \times 10^{-2}$, formally an order of magnitude above the canonical ISM value of 10^{-2} [76]. However, if the unmodeled neutral and molecular gas were included, the true δ_{mw} would likely decrease toward ISM-like values. Thus, the high apparent dust-to-gas ratios primarily trace modeling geometry and limited gas coverage rather than an actual overabundance of solids.

Object-by-object trends (**Tables 5–9**) reveal that dustier nebulae exhibit strong refractory depletions (Mg, Si, Fe) and mixed carbonaceous–silicate compositions, while dust-poor objects show weak depletions and low grain abundances, reflecting

PN G	δ	M_{tot} ($M_\odot$)	M_{dust} ($M_\odot$)	f_{dust}	Grain abn (log)
019.8 + 05.6	1.611e + 00	9.854e – 05	6.080e – 05	0.617	2.29[a]
021.2 + 02.9	2.631e – 02	1.016e – 03	2.604e – 05	0.026	0.62[b]
021.3 + 02.2	6.333e – 02	1.161e – 03	6.915e – 05	0.060	0.92[a]
034.5 – 11.7	3.694e – 01	1.267e – 02	3.418e – 03	0.270	1.63[a]
048.5 + 04.2	1.359e – 02	7.391e – 03	9.910e – 05	0.013	0.50[a]
048.8 + 08.4	1.050e – 02	8.348e – 05	8.674e – 07	0.010	0.52[b]
050.4 – 01.6	3.201e – 02	2.252e – 03	6.985e – 05	0.031	0.69[a]
059.8 – 00.5	8.407e – 02	7.091e – 03	5.499e – 04	0.078	1.43[b]
063.0 + 00.5	6.935e – 02	5.431e – 03	3.522e – 04	0.065	1.38[b]
080.2 + 03.8	1.612e – 03	2.841e – 03	4.572e – 06	0.002	–0.55[a]
095.2 + 25.4	1.101e + 00	5.178e – 06	2.713e – 06	0.524	2.21[a]
135.9 + 55.9	4.780e – 03	2.086e – 05	9.924e – 08	0.005	1.66[a]
153.7 – 01.4	4.981e – 02	3.394e – 02	1.610e – 03	0.047	0.92[a]
183.0 + 00.0	8.859e – 02	2.197e – 02	1.788e – 03	0.081	1.13[a]
202.9 + 07.4	1.881e – 02	1.166e – 04	2.153e – 06	0.018	0.90[a]
216.9 – 05.2	1.185e – 02	3.861e – 04	4.522e – 06	0.012	0.77[a]

Notes: δ represents the dust-to-gas mass ratio; f_{dust} is the fraction of total mass in dust. All masses are expressed in solar units ($M_\odot$). Grain abundances are given in logarithmic scale.

Table 5.
Gas and dust parameters for the planetary nebula (PN) sample. Superscripts [a] and [b] on grain abundances indicate values adopted from [112] and [113], respectively.

inefficient condensation or post-AGB dust erosion caused by photo-processing and shocks [114–116]. This diversity underscores that the dust content of PNe is shaped not only by progenitor C/O chemistry but also by the efficiency of dust survival within the ionized envelope [117, 118].

In a broader Galactic context, PNe thus act as modest but compositionally significant contributors to ISM enrichment. They inject chemically diverse refractory grains and metal-rich gas, but their total mass return is relatively small $\sim 10^{-3} - 10^{-4} M_\odot$ of dust per nebula). Our results, therefore, support a picture in which planetary nebulae provide *locally enriched yet globally minor* inputs to the Galactic dust cycle – supplying the ISM with a heterogeneous mixture of partially processed solids and ionized metals, whose long-term fate depends on subsequent interstellar reprocessing and grain survival [119].

6.1 Discussion of outliers in our pne sample

PN G 080.2 + 03.8, an O–rich and dust–poor nebula, displays an atypical depletion pattern: very low C and O depletions but high Mg and Fe, with moderate Si. From the combined gas + grain abundances (**Table 8**), the inferred solid-phase fractions are $\sim$9% for C, 0.4% for O, 85% for Mg, 48% for Si, and 95% for Fe. Its overall dust content is among the lowest in the sample, with $\delta \approx 1.6 \times 10^{-3}$, $f_{dust} \approx 0.002$, and a total $M_{dust} = 4.6 \times 10^{-6} M_\odot$ (**Tables 5–8**).

PN G	He	N	O	Ne	Ar	S	Cl
			$[X/H]_{gas}$ (dex)				
019.8 + 05.6	- 0.74	0.57	0.26	- 0.35	- 0.85	0.78	–
021.2 + 02.9	–	–	0.26	–	- 0.35	–	–
021.3 + 02.2	- 0.30	- 0.69	- 0.07	–	- 0.15	- 0.63	–
034.5 – 11.7	- 1.35	- 0.20	- 0.60	–	–	–	- 1.39
048.5 + 04.2	0.49	- 0.89	- 0.46	- 0.79	- 0.34	–	–
048.8 + 08.4	0.71	1.52	- 0.53	- 0.21	- 0.79	–	- 0.69
050.4 – 01.6	0.02	- 0.71	- 0.42	–	- 1.26	- 0.96	–
059.8 – 00.5	0.70	1.47	1.09	–	0.30	0.43	–
063.0 + 00.5	0.78	–	0.74	–	- 0.45	–	–
080.2 + 03.8	–	–	1.33	–	0.08	–	–
095.2 + 25.4	- 1.59	- 0.35	- 0.26	2.23	0.61	0.48	–
135.9 + 55.9	2.39	–	–	–	–	–	–
153.7 – 01.4	0.14	0.01	- 0.25	–	- 0.44	- 0.23	–
183.0 + 00.0	- 0.01	- 1.89	- 0.32	- 0.18	- 0.74	–	–
202.9 + 07.4	0.92	–	- 0.34	–	–	–	–
216.9 – 05.2	1.01	–	- 0.65	- 1.22	0.17	–	–

Solar reference values are listed in the caption; dashes denote unavailable abundances.

Table 6.
Gas-phase depletions for the PN sample. Entries give $[X/H]_{gas}$ in dex, computed from the provided $12 + \log_{10}(X/H)$ values minus the adopted solar baseline (He = 10.93, N = 7.83, O = 8.69, Ne = 7.93, Ar = 6.38, S = 7.12, Cl = 5.50).

	C	O	Mg	Si	Fe
Mass fraction (%)	37.2	23.3	8.9	10.2	20.4

Fractions are by mass and sum to 100%. A convenient grouping for compact reporting is C = 37.2% and silicates/oxides (O + Mg + Si + Fe) = 62.8%.

Table 7.
Mass fractions of the grain mixture inferred from the cloudy grain chemical composition logs (C, O, Mg, Si, Fe per h in grains).

The exceptionally low oxygen depletion stems from its large gas-phase O reservoir ($\log[O/H] \approx -1.98$; **Table 6**), which dilutes the fraction of O bound in solids. At the same time, the preferential survival of refractory grains enhances Fe and Mg relative to Si and C. Laboratory and theoretical studies indicate that Si–O networks are more easily sputtered or photodesorbed than Fe- and Mg-bearing phases, leaving behind a refractory-rich dust residue embedded in an O-dominated gas.

The extremely low carbon depletion aligns with an O-rich (C/O < 1) outflow, where carbon dust formation is suppressed – carbon remains bound in CO or in the gas phase, and small carbonaceous grains or PAHs are easily destroyed in a hard radiation field [58]. Overall, PN G 080.2 + 03.8 contains a silicate-processed,

PN G	$M_{dust,C}$	$M_{dust,O}$	$M_{dust,Mg}$	$M_{dust,Si}$	$M_{dust,Fe}$
	$(M_\odot)$	$(M_\odot)$	$(M_\odot)$	$(M_\odot)$	$(M_\odot)$
019.8 + 05.6	2.260e − 05	1.419e − 05	5.390e − 06	6.229e − 06	1.239e − 05
021.3 + 02.2	2.57e − 05	1.61e − 05	6.16e − 06	7.08e − 06	1.41e − 05
021.2 + 02.9	9.69e − 06	6.06e − 06	2.32e − 06	2.68e − 06	5.31e − 06
034.5 − 11.7	1.27e − 03	7.96e − 04	3.04e − 04	3.52e − 04	6.97e − 04
048.5 + 04.2	3.69e − 05	2.31e − 05	8.82e − 06	1.02e − 05	2.02e − 05
048.8 + 08.4	3.23e − 07	2.02e − 07	7.71e − 08	8.94e − 08	1.77e − 07
050.4 − 01.6	2.60e − 05	1.62e − 05	6.22e − 06	7.20e − 06	1.43e − 05
050.4 − 01.6	2.598e − 05	1.628e − 05	6.217e − 06	7.195e − 06	1.425e − 05
059.8 − 00.5	2.040e − 04	1.280e − 04	4.870e − 05	5.630e − 05	1.120e − 04
063.0 + 00.5	1.31e − 04	8.21e − 05	3.13e − 05	3.63e − 05	7.18e − 05
080.2 + 03.8	1.700e − 06	1.065e − 06	4.070e − 07	4.710e − 07	9.320e − 07
095.2 + 25.4	1.009e − 06	6.321e − 07	2.414e − 07	2.523e − 07	5.541e − 07
135.9 + 55.9	3.691e − 08	2.312e − 08	8.833e − 09	1.021e − 08	2.025e − 08
153.7 − 01.4	5.996e − 04	3.761e − 04	1.390e − 04	1.605e − 04	3.293e − 04
183.0 + 00.0	6.647e − 04	4.168e − 04	1.562e − 04	1.830e − 04	3.755e − 04
202.9 + 07.4	8.015e − 07	5.025e − 07	1.883e − 07	2.206e − 07	4.525e − 07
216.9 − 05.2	1.68e − 06	1.05e − 06	4.02e − 07	4.66e − 07	9.23e − 07

All masses are in solar units ($M_\odot$). Per-element entries provide the contribution of each element to the total dust mass, derived from the CLOUDY "Grain Chemical Composition" number abundances and atomic weights.

Table 8.
Element-resolved dust mass budget per PN. Filled rows are based on cloudy outputs; other rows are placeholders to be completed as models are processed.

refractory-rich dust residue (Fe and Mg predominantly in solids, Si only partially depleted) embedded in an oxygen-dominated gas. It therefore stands as a clear outlier compared with the rest of the sample, which typically shows 40–60% oxygen depletion (**Table 9**), and likely returns most of its metals to the ISM as gas-phase oxygen with a minor contribution from Fe/Mg-rich grains.

PN G 019.8 + 05.6 This nebula is an extreme, dust-dominated outlier within the sample. Its dust-to-gas mass ratio is exceptionally high ($\delta \approx 1.6$; **Table 5**), corresponding to nearly two orders of magnitude above the canonical ISM value. The dust mass fraction is likewise large ($f_{dust} \approx 0.62$), indicating that the majority of the total nebular mass is contained in solids rather than gas. The total dust mass derived from the photoionization model is $M_{dust} = 6.1 \times 10^{-5} M_\odot$, among the largest in the sample (**Table 8**).

Element-resolved analysis reveals a mixed carbonaceous–silicate composition, with approximate mass contributions of C:O:Mg:Si:Fe = 37:23:9:10:20% (**Table 7**). Depletion diagnostics confirm that nearly all Mg, Si, and Fe, and the majority of C and O, are locked into grains – depletion fractions exceeding 95% for most refractory species (**Table 9**). Such an elevated δ most likely reflects highly efficient dust condensation during the late AGB phase and minimal subsequent grain destruction. Consequently, the material returned by PN G 019.8 + 05.6 to the ISM is

PN G	C (%)	O (%)	Mg (%)	Si (%)	Fe (%)
019.8 + 05.6	98.6	96.7	~100	99.8	~100
021.2 + 02.9	43.3	65.7	98.8	98.8	99.8
021.3 + 02.2	74.9	72.9	98.3	96.1	99.6
034.5 – 11.7	98.6	95.0	99.9	99.9	99.9
048.5 + 04.2	50.4	72.7	97.0	97.0	99.5
048.8 + 08.4	49.6	77.5	98.9	98.9	99.6
050.4 – 01.6	88.8	70.8	98.4	98.4	99.6
063.0 + 00.5	94.9	97.5	99.2	99.2	99.9
080.2 + 03.8	9.2	0.4	85.4	48.3	94.9
095.2 + 25.4	95.1	97.1	99.7	99.7	99.7
135.9 + 55.9	86.3	86.9	99.4	99.4	99.4
153.7 – 01.4	77.4	42.8	99.5	97.2	99.7
183.0 + 00.0	81.8	56.6	99.6	97.8	99.8
202.9 + 07.4	76.3	39.8	99.3	96.8	99.7
216.9 – 05.2	81.2	77.9	99.6	99.6	99.9

Depletion fraction = number of atoms in grains/(gas + grains)

Table 9.
Elemental depletions into dust for the PN sample. Values are the fraction of each element locked in the solid phase, computed from gas + grain abundances.

dominated by solids – both silicate/oxide and carbonaceous components – with only a minor fraction of metals remaining in the gas phase (see **Table 6** for gas-phase abundances).

PN G 095.2 + 25.4 Like PN G 019.8 + 05.6, PN G 095.2 + 25.4 is also dust-rich in a *fractional* sense, with $\delta \approx 1.1$ ($\sim 10^2$ times the canonical ISM value) and $f_{\text{dust}} \approx 0.52$ (**Table 5**). However, its absolute reservoir is small, with a total mass of only a few $10^{-6} M_\odot$ and a dust mass $M_{\text{dust}} = 2.7 \times 10^{-6} M_\odot$ (**Table 8**). The element-resolved dust composition and high grain abundances inferred from the cloudy grain logs indicate strong depletion of refractory species into solids, forming a broadly mixed carbonaceous and silicate population (**Tables 7** and **9**). In practice, PN G 095.2 + 25.4 returns a *dust-biased* payload to the ISM, but the overall mass injection remains modest because of its intrinsically low nebular mass. The combination of large δ and small M_{tot} may reflect either efficient condensation in a low-mass envelope or an underestimate of the total gas mass due to missing neutral or molecular components. In either case, the internal dust–gas balance across our tables remains self-consistent, pointing to refractory-rich grains as the dominant metal reservoir.

PN G 135.9 + 55.9 In sharp contrast, PN G 135.9 + 55.9 is *dust-poor*, with $\delta \approx 4.8 \times 10^{-3}$ (~ 0.5 times the ISM value) and $f_{\text{dust}} \approx 0.005$ (**Table 5**). Despite grain-number abundances suggestive of a refractory mixture, the total dust mass is negligible ($M_{\text{dust}} = 9.9 \times 10^{-8} M_\odot$; **Table 8**). This pattern likely results from inefficient dust formation, such as carbon-poor and oxygen-rich ejecta at low metallicity, or from substantial post-AGB grain destruction through thermal sputtering or photo-desorption in a hard radiation field. These processes preferentially erode silicate networks and small carbonaceous particles, leaving only the most robust refractory phases. Consequently, the ISM return from PN G 135.9 + 55.9 is dominated by gas-phase metals, with only a minor contribution from surviving dust grains.

7. Conclusions

The analysis of 16 PNe provides a quantitative picture of their gas and dust budgets and allows direct comparison with interstellar dust depletion patterns. A key conclusion is that *absolute yields matter*: a high δ does not necessarily translate into a large absolute dust mass. Several nebulae in our sample exhibit high fractional dust content yet are of intrinsically low mass, implying that their contribution to the Galactic dust reservoir is limited despite their apparent dust richness. This agrees with modeling studies of individual PNe, where δ values of order 10^{-3}–10^{-4} are typical [65, 68, 120, 121]. Our derived mean $\langle\delta\rangle_{\rm samp} \approx 0.09$ should be interpreted as characteristic of the photoionized regions modeled here, not of the full nebular or molecular envelopes that would increase the total gas mass and proportionally reduce δ.

A second outcome is that *refractory selectivity is common*. In systems where grains survive, magnesium and iron are consistently the most depleted species, in agreement with their incorporation into highly refractory silicate and oxide phases. Silicon depletion tends to be moderate, while carbon depletion can be weak or absent in oxygen-rich nebulae. This pattern is consistent with previous observational and theoretical studies of dust condensation and destruction in evolved stars and PNe, which highlight the durability of Fe/Mg-rich components and the vulnerability of carbonaceous and silicate networks to photo-processing and sputtering [58, 77].

Relative to the canonical interstellar value of $\delta_{\rm ISM} \sim 0.01$ [76, 122], the mean ratio in our sample ($\langle\delta\rangle_{\rm samp} \sim 0.09$) is higher. This outcome is expected, since compact PNe represent transient, metal-enriched outflows rather than well-mixed ISM gas. The object-to-object diversity, set by progenitor C/O chemistry, condensation efficiency, and subsequent photo- or shock-processing, determines whether metals are ultimately returned to the ISM as gas or as grains. In this sense, our PNe bridge the chemical evolution between stellar mass loss and interstellar enrichment. Chemically processed solids can persist through ISM cycling. These grains provide durable seeds for ice accretion and molecular growth in dense clouds, establishing a clear physical and chemical continuity between the terminal stages of stellar evolution and the raw materials from which new stars and planetary systems form.

Nomenclature

AAO	Anglo-Australian Observatory
AGB	Asymptotic Giant Branch
AIB	Aromatic Infrared Band
CCD	Charge-Coupled Device, optical imaging detector
CLOUDY	Photoionization modeling code by G. Ferland et al.
CS	Central Star
CSPN	Central Star of a Planetary Nebula

HASH	Hong Kong/AAO/Strasbourg H_α Planetary Nebula Database
HAC	Hydrogenated Amorphous Carbon
H–R	diagram Hertzsprung–Russell diagram, luminosity–temperature relation
IR	Infrared
IRAS	Infrared Astronomical Satellite
ISM	Interstellar Medium
JWST	James Webb Space Telescope
ISO	Infrared Space Observatory
MIR	Mid-Infrared
NEAT	NEBULAR EMPIRICAL ANALYSIS TOOL
PAH	Polycyclic Aromatic Hydrocarbon
PDR	Photodissociation Region
PN	PNe Planetary Nebula, Planetary Nebulae
RGB	Composite image convention; Red: W4, Green: W3, Blue: W2
SED	Spectral Energy Distribution
UV	Ultraviolet
WISE	Wide-field Infrared Survey Explorer

Symbols and greek letters

α	Alpha – used for α-element notation (e.g., O, Mg, Si, S, Ca)
δ	Dust-to-gas mass ratio ($M_{\mathrm{dust}}/M_{\mathrm{gas}}$)
δ_{samp}	Sample-averaged dust-to-gas ratio
δ_{mw}	Modeled dust-to-gas ratio within the ionized region only
$\langle\delta\rangle_{\mathrm{samp}}$	Mean dust-to-gas ratio across the sample
ℓ_X	Logarithmic grain number abundance for element or species X
λ	Wavelength
μm	Micrometer; 10^{-6} m (micron)
$\rho_{\mathrm{d}}/\rho_{\mathrm{g}}$	Dust-to-gas mass density ratio
X/H	Logarithmic abundance ratio of element X relative to the Sun
amu, m_{u}	Atomic mass unit
A_V	Visual extinction
A_X	Atomic weight of element; Xf_{dust} Fraction of total mass in dust
GHz	Gigahertz; unit of frequency (10^9 Hz)
H α	Hydrogen-alpha emission line (6562.80 Å)
H β	Hydrogen-beta emission line (4861.33 Å)
Jy	Jansky; unit of flux density (10^{-26} W m^{-2} Hz^{-1})
K	Kelvin; unit of temperature
m_X, m_Y	Dust-phase masses of elements X, Y
$M_\odot$	Solar mass; 1.989×10^{30} kg
M_{dust}	Total dust mass
M_{gas}	Total gas mass
M_{tot}	Total nebular mass ($M_{\mathrm{gas}} + M_{\mathrm{dust}}$)
n_X,n_Y	Grain-phase number abundances of elements X, Y
ppb	Parts per billion
sp^2, sp^3	Carbon hybridization types in amorphous or graphitic solids
Al_2O_3	Corundum (Aluminum Oxide)

C_2	Diatomic carbon molecule
CO	Carbon Monoxide
CO_2	Carbon Dioxide
C_{60}, C_{70}	Fullerene having 60, 70 carbons
CH_3	OH Methanol
CN	Cyanogen radical
H_2	Molecular Hydrogen
$MgAl_2O_4$	Spinel (Magnesium Aluminate)
MgS	Magnesium Sulfide
Mg_2SiO_4	Forsterite, magnesium-rich silicate mineral
$MgSiO_3$	Enstatite, magnesium silicate mineral
NH_2	Amidogen radical
OH	Hydroxyl Radical
SiC	Silicon Carbide
Si_3N_4	Silicon Nitride

Author details

Şengül Yalgın
Graduate School of Natural and Applied Sciences, Ankara University, Ankara, Türkiye

*Address all correspondence to: sengulyalgin@gmail.com

References

[1] Abell GO. Properties of some old planetary nebulae. The Astrophysical Journal. 1966;**144**:259. DOI: 10.1086/148602

[2] Karakas AI, Lattanzio JC. The Dawes review 2: Nucleosynthesis and stellar yields of low- and intermediate-mass single stars. Publications of the Astronomical Society of Australia. 2014;31. DOI: 10.1017/pasa.2014.21

[3] Tosi S. Insight on AGB mass-loss and dust production from PNe. Galaxies. 2024;**12**:85. DOI: 10.3390/galaxies12060085

[4] Pottasch SR. Planetary Nebulae – A Study of Late Stages of Stellar Evolution. In: Sky & Telescope. 1984;**67**(6):527

[5] Messier C. Catalogue des nébuleuses et des amas d'étoiles. Connoissance Des Temps Ou Des Mouvements Célestes, pour I"annee 1784. 1781;227–267

[6] Huggins W. On the spectra of some of the chemical elements. Philosophical Transactions of the Royal Society of London Series I. 1864;**154**:139–160

[7] Hubble E. No. 241. A general study of diffuse galactic nebula. Contributions from the Mount Wilson Observatory/Carnegie Institution of Washington. 1922;**241**:1–38

[8] Menzel DH. The planetary nebulae. Publications of the Astronomical Society of the Pacific. 1926;**38**:295. DOI: 10.1086/123616

[9] Zanstra H. Temperatures of stars in planetary nebulæ. Nature. 1928;**121**:790–791. DOI: 10.1038/121790a0

[10] Bowen IS. The origin of the nebular lines and the structure of the planetary nebulae. The Astrophysical Journal. 1928;**67**:1–15. DOI: 10.1086/143091

[11] Shklovsky IS. The nature of planetary nebulae and their nuclei. Astronomicheskii Zhurnal. 1956;**33**:315–329

[12] Paczyṅski B. Evolution of single stars. VI. Model nuclei of planetary nebulae. Acta Astronomica. 1971;**21**:417

[13] Barker T. Spectrophotometry of planetary nebulae. II. Chemical abundances. The Astrophysical Journal. 1978;**220**:193–209. DOI: 10.1086/155896

[14] Kwok S. From red giants to planetary nebulae. The Astrophysical Journal. 1982;**258**:280–288. DOI: 10.1086/160078

[15] Ferland GJ 1983. Hazy, A Brief Introduction to Cloudy 33. University of Kentucky Internal Report, 565

[16] Oey MS, Shields J, Dopita M, Smith R. Calibrating nebular diagnostics of T* and abundance. Stars, Gas and Dust in Galaxies: Exploring the Links. 2000;**221**:79

[17] Dopita MA, Sutherland RS. The importance of photoelectric heating by dust in planetary nebulae. The Astrophysical Journal. 2000;**539**:742–750. DOI: 10.1086/309241

[18] Kwok S, Dopita M, Sutherland R. Planetary nebulae: Their evolution and role in the universe. Cambridge, United Kingdom Cambridge University Press. 2003

[19] Soker N, Kastner JH. Magnetic flares on asymptotic giant branch stars. The Astrophysical Journal. 2003;**592**:498–503. DOI: 10.1086/375686

[20] van Winckel H, Reyniers M. Tracing AGV s-process nucleo-synthesis by the analysis of post-AGB stars. Astronomische Nachrichten. 2003;**324**:20

[21] Magrini L, Perinotto M, Mampaso A, Corradi RLM. Chemical abundances of planetary nebulae in M 33. Astronomy and Astrophysics. 2004;**426**:779–786. DOI: 10.1051/0004-6361:20041103

[22] De Marco O. Planetary Nebulae, Globular Clusters and Binary Mergers Baltimore, MD, USA. Space Telescope Science Institute (STScI), HST Proposal; 2009

[23] Frew DJ, Parker QA. Planetary nebulae: Observational properties, mimics and diagnostics. Publications of the Astronomical Society of Australia. 2010;**27**:129–148. DOI: 10.1071/AS09040

[24] Cohen M, Parker QA, Green AJ, Miszalski B, Frew D, Murphy T. Multiwavelength diagnostic properties of Galactic planetary nebulae detected by the GLIMPSE-I. Monthly Notices of the Royal Astronomical Society. 2011;**413**:514–542. DOI: 10.1111/j.1365-2966.2010.18157.x

[25] Wesson R, Stock DJ, Scicluna P. Understanding and reducing statistical uncertainties in nebular abundance determinations. Monthly Notices of the Royal Astronomical Society. 2012;**422**:3516–3526. DOI: 10.1111/j.1365-2966.2012.20863.x

[26] Zijlstra AA. Planetary nebulae in 2014: A review of research. Revista Mexicana de Astronomia Y Astrofisica. 2015;**51**:221. DOI: 10.48550/arXiv.1506.05508

[27] Delgado-Inglada G, Morisset C, Sánchez S. Ionization correction factors for planetary. 2014;**440**(1). DOI: 10.1093/mnras/stu341

[28] Miller Bertolami MM. New models for the evolution of post-asymptotic giant branch stars and central stars of planetary nebulae. Astronomy and Astrophysics. 2016;588. DOI: 10.1051/0004-6361/201526577

[29] Miller Bertolami MM. Evolutionary timescales from the AGB to the CSPNe phase. Why Galaxies Care about AGB Stars: A Continuing Challenge through Cosmic Time. 2019;**343**:36–46. DOI: 10.1017/S1743921318007330

[30] Parker QA, Bojicic IS, Frew DJ 2016. HASH: The Hong Kong/AAO/Strasbourg Hα planetary nebula database. Journal of Physics Conference Series 728. DOI: 10.1088/1742-6596/728/3/032008

[31] Fragkou V, Bojicic I, Frew D, Parker Q. Planetary nebula candidates uncovered with the HASH research platform. Planetary Nebulae: Multi-Wavelength Probes of Stellar and Galactic Evolution. 2017;**323**:329–330. DOI: 10.1017/S1743921317000473

[32] Bojicic IS, Parker QA, Frew DJ. The Hong Kong/AAO/Strasbourg Hα (HASH) planetary nebula database.

Planetary Nebulae: Multi-Wavelength Probes of Stellar and Galactic Evolution. 2017;**323**:327–328. DOI: 10.1017/S1743921317003234

[33] Le Dû P, et.al. Amateur PN discoveries and their spectral confirmation: A significant new addition to the Galactic PN inventory. Astronomy and Astrophysics. 2022:666. DOI: 10.1051/0004-6361/202243393

[34] Ritter A, Parker QA, Lykou F, Zijlstra AA, Guerrero MA, Le Du P. From an amateur PN candidate to the Rosetta Stone of SN Iax research. arXiv E-prints. 2023. DOI: 10.48550/arXiv.2311.03700

[35] Peimbert M, Peimbert A, Delgado-Inglada G. Nebular spectroscopy: A guide on Hii regions and planetary nebulae. Publications of the Astronomical Society of the Pacific. 2017;**129**:082001. DOI: 10.1088/1538-3873/aa72c3

[36] Parker QA. Planetary nebulae and how to find them: A concise review. Frontiers in Astronomy and Space Sciences. 2022;**9**. DOI: 10.3389/fspas.2022.895287

[37] Kwitter KB, Henry RBC. Planetary nebulae: Sources of enlightenment. Publications of the Astronomical Society of the Pacific. 2022;134. DOI: 10.1088/1538-3873/ac32b1

[38] Chatzikos M, et.al The 2023 release of Cloudy. Revista Mexicana de Astronomía Y Astrofísica. 2023;**59**:327–343

[39] Kwok S. Planetary nebulae research: Past, present, and future. Galaxies. 2024;**12**:39. DOI: 10.3390/galaxies12040039

[40] Zhang CY, Kwok S. 179th Meeting of the American Astronomical Society Washington, DC, USA. Bulletin of the American Astronomical Society; 1991;**23**:1407

[41] Bilikova J, Chu Y-H, Su KYL, Gruendl R. Spectral analysis of mid-IR excesses of central stars of planetary nebulae. Planetary Nebulae: An Eye to the Future. 2012;**283**:310–311. DOI: 10.1017/S1743921312011179

[42] Otsuka M, et.al. XSHOOTER spectroscopy of the enigmatic PN Lin49 in the SMC. Planetary Nebulae: Multi-Wavelength Probes of Stellar and Galactic Evolution. 2017;**323**:254–258

[43] Willenbrink KA, Hrivnak B, Volk K, Su KYL. Spectroscopic classification of IR-excess IRAS sources. American Astronomical Society Meeting Abstracts. 2018;**231**:146.10

[44] Goebel JH, Moseley SH. MgS grain component in circumstellar shells. The Astrophysical Journal. 1985;**290**:L35–L39. DOI: 10.1086/184437

[45] Volk K, Sloan GC, Kraemer KE. The 21 μm and 30 μm emission features in carbon-rich objects. Astrophysics and Space Science. 2020;**365**. DOI: 10.1007/s10509-020-03798-2

[46] Treffers R, Cohen M. High-resolution spectra of cool stars in the 10- and 20-micron regions. The Astrophysical Journal. 1974;**188**:545–552. DOI: 10.1086/152746

[47] Speck AK, Whittington AG. Collaborative research: An integrated experimental and observational study of cosmic silicate astromineralogy. NSF Award. 2009;**09**:8302

[48] Cami J, Bernard-Salas J, Peeters E, Malek SE. Detection of C_{60} and C_{70} in a young planetary nebula. Science. 2010;**329**:1180. DOI: 10.1126/science.1192035

[49] Garcia-Hernández DA, et.al. Formation of fullerenes in H-containing planetary nebulae. The Astrophysical Journal. 2010;**724**:L39–L43

[50] Jones AP, Duley WW, Williams DA. The structure and evolution of hydrogenated amorphous carbon grains and mantles in the interstellar medium. Quarterly Journal of the Royal Astronomical Society. 1990;**31**:567–582

[51] Scott GBI 1997. Experimental Studies of Astrochemical Ion-atom and Ion-molecule Reactions. Ph.D. Thesis

[52] Sloan GC, et.al. The unusual hydrocarbon emission from the early carbon star HD 100764: The connection between aromatics and aliphatics. The Astrophysical Journal. 2007;**664**: 1144–1153

[53] Draine BT. Grain destruction in interstellar shock waves. Astrophysics and Space Science. 1995;**233**:111–123. DOI: 10.1007/BF00627339

[54] Waters LBFM, et.al. Crystalline silicates in planetary nebulae with [WC] central stars. Astronomy and Astrophysics. 1998;**331**:L61–L64

[55] Molster FJ, Waters LBFM, Tielens AGGM, Barlow MJ. VizieR Online Data Catalog: Crystalline Silicates around Evolved Stars (Molster +, 2002). 2001;**338**:J/A+A/382/184. DOI: 10.26093/cds/vizier.33820184

[56] Gielen C et.al. 2011. Dusty discs in Galactic and extragalactic post-AGB binaries. Asymmetric Planetary Nebulae 5 Conference.

[57] Allamandola LJ, Tielens AGGM 1989. Interstellar dust: proceedings of the 135th Symposium of the International Astronomical Union. Santa Clara, California, 26-30 July 1988. Interstellar Dust 135.

[58] Tielens AGGM. Interstellar polycyclic aromatic hydrocarbon molecules. Annual Review of Astronomy and Astrophysics. 2008;**46**:289–337. DOI: 10.1146/annurev.astro.46.060407.145211

[59] Kastner J, et.al. JWST/NIRCam imaging of the bipolar planetary nebula NGC 6537: The (infra)red spider, revealed. arXiv E-prints. 2025

[60] Matsuura M, et.al. The JWST/MIRI view of the planetary nebula NGC 6302 - I. A UV-irradiated torus and a hot bubble triggering PAH formation. Monthly Notices of the Royal Astronomical Society. 2025;**542**: 1287–1307

[61] Clark N, et.al. JWST observations of the Ring Nebula (NGC 6720) – II. PAH emission. Monthly Notices of the Royal Astronomical Society. 2025;**540**:1984–1997

[62] Ventura P, Limongi M. Gas and dust from stars. Memorie Della Societa Astronomica Italiana. 2014;**85**:196

[63] Stephenson M, et al. Heavy element abundances in Magellanic Cloud planetary nebulae. American Astronomical Society Meeting Abstracts 2022;**240**:351.03

[64] Boothroyd AI, Sackmann I-J, Ahern SC. Prevention of high-luminosity carbon stars by hot bottom burning. The Astrophysical Journal. 1993;**416**:762. DOI: 10.1086/173275

[65] Dell'Agli F, et.al. Dust from evolved stars: A pilot analysis of the AGB to PN transition. Monthly Notices of the Royal Astronomical Society. 2023;**526**:5386–5392

[66] Tosi S, et.al. Planetary nebulae of the Large Magellanic Cloud. I. A multiwavelength analysis. Astronomy and Astrophysics. 2024:688. DOI: 10.1051/0004-6361/202449588

[67] Tosi S, Marini E. Tracing the evolution of the emission properties of carbon-rich AGB, post-AGB, and PN sources. Astronomy. 2025;**4**:2. DOI: 10.3390/astronomy4010002

[68] Ventura P, et.al. Planetary nebulae of the Large Magellanic Cloud: II. The connection with the progenitors' properties. Astronomy and Astrophysics. 2025:694. DOI: 10.1051/0004-6361/202452188

[69] Ferrarotti AS, Gail H-P. Composition and quantities of dust produced by AGB-stars and returned to the interstellar medium. Astronomy and Astrophysics. 2006;**447**:553–576. DOI: 10.1051/0004-6361:20041198

[70] Jenkins EB. A unified representation of gas-phase element depletions in the interstellar medium. The Astrophysical Journal. 2009;**700**:1299–1348. DOI: 10.1088/0004-637X/700/2/1299

[71] Hensley BS, Draine BT, Meisner AM. A case against spinning PAHs as the source of the anomalous microwave emission. The Astrophysical Journal. 2016;**827**:45. DOI: 10.3847/0004-637X/827/1/45

[72] Compiègne M, et.al. The global dust SED: Tracing the nature and evolution of dust with DustEM. Astronomy and Astrophysics. 2011;**525**. DOI: 10.1051/0004-6361/201015292

[73] Delgado-Inglada G, Mesa-Delgado A, Garcia-Rojas J, Rodriguez M, Esteban C. The Fe/Ni ratio in ionized nebulae: Clues on dust depletion patterns. Monthly Notices of the Royal Astronomical Society. 2016;**456**:3855–3865. DOI: 10.1093/mnras/stv2961

[74] Jones AP, Köhler M, Ysard N, Bocchio M, Verstraete L. The global dust modelling framework THEMIS. Astronomy and Astrophysics. 2017;602. DOI: 10.1051/0004-6361/201630225

[75] Hensley BS, Draine BT. Observational constraints on the physical properties of interstellar dust in the post-Planck era. The Astrophysical Journal. 2021;**906**:73. DOI: 10.3847/1538-4357/abc8f1

[76] Draine BT. Physics of the Interstellar and Intergalactic Medium. Princeton, NJ: Published by Princeton University Press; 2011

[77] Jones AP. Interstellar dust: interfacing laboratory, theoretical and observational studies Cambridge, United Kingdom. IAU General Assembly; 2015. p. 2256693

[78] Wolff AJ 2006. Adsorption and Desorption of Mixed Molecular Ices from a Cosmic Dust Grain Analogue Surface. Ph.D. Thesis.

[79] Clayton DD, Nittler LR. Astrophysics with presolar stardust.

Annual Review of Astronomy and Astrophysics. 2004;**42**:39–78. DOI: 10.1146/annurev.astro.42.053102.134022

[80] van Dishoeck EF, et.al. JWST observations of young protostars (JOYS): Overview of program and early results. Astronomy and Astrophysics. 2025:699. DOI: 10.1051/0004-6361/202554444

[81] Lee D-H, et.al. Diffuse far-ultraviolet observations of the Taurus region. The Astrophysical Journal. 2006;**644**:L181–L184

[82] Paneque-Carreño T, Miotello A, van Dishoeck EF, Rosotti G, Tabone B. Vertical CO surfaces as a probe for protoplanetary disk mass and carbon depletion. Astronomy and Astrophysics. 2025;698. DOI: 10.1051/0004-6361/202451862

[83] van Dishoeck EF, Bergin EA. Astrochemistry associated with planet formation. arXiv E-prints. 2020

[84] Oberg N, Kamp I, Cazaux S, Woitke P, Thi WF. Circumplanetary disk ices. I. Ice formation vs. viscous evolution and grain drift. Astronomy and Astrophysics. 2022;667. DOI: 10.1051/0004-6361/202244092

[85] van Dishoeck EF, et.al. The diverse chemistry of protoplanetary disks as revealed by JWST. Faraday Discussions. 2023;**245**:52–79

[86] van Dishoeck E 2023. The birth of stars and planets: New views from ALMA and JWST. 3rd Annual Conference of the African Astronomical Society (AfAS-2023)

[87] Amari S, Zinner E, Gallino R. Abundances of presolar graphite and SiC from supernovae and AGB stars in the Murchison meteorite. Origin of Matter and Evolution of Galaxies. 2013;**1594,**:307–312. DOI: 10.1063/1.4874087

[88] Nittler LR, Ciesla F. Astrophysics with extraterrestrial materials. Annual Review of Astronomy and Astrophysics. 2016;**54**:53–93. DOI: 10.1146/annurev-astro-082214-122505

[89] Nittler LR, Alexander CMO. Automated isotopic measurements of micron-sized dust: Application to meteoritic presolar silicon carbide. Geochimica Et Cosmochimica Acta. 2003;**67**:4961–4980. DOI: 10.1016/S0016-7037(03)00485-X

[90] Hoppe P, Strebel R, Eberhardt P, Amari S, Lewis RS. Isotopic properties of silicon carbide X grains from the Murchison meteorite in the size range 0.5–1.5 μm. Meteoritics and Planetary Science. 2000;**35**:1157–1176. DOI: 10.1111/j.1945-5100.2000.tb01505.x

[91] Amari S, Zinner E, Gallino R. Presolar graphite from the Murchison meteorite: An isotopic study. Geochimica Et Cosmochimica Acta. 2014;**133**:479–522. DOI: 10.1016/j.gca.2014.01.006

[92] Coria D. Exoplanetary origins: Unraveling planetary formation and accretion histories with CNO isotopologues. Cool Stars. 2024;**22**. DOI: 10.5281/zenodo.13619959

[93] Lugaro M, Karakas AI, Petö M, Plachy E. Do meteoritic silicon carbide grains originate from asymptotic giant branch stars of super-solar metallicity? Geochimica Et Cosmochimica Acta. 2018;**221**:6–20. DOI: 10.1016/j.gca.2017.06.006

[94] Reach WT, et.al. Composition of planetary debris around the white dwarf GD 362. arXiv E-prints. 2025

[95] Nittler LR, et.al A cometary building block in a primitive asteroidal meteorite. Nature Astronomy. 2019;**3**:659–666

[96] Leitner J, Hoppe P 2022. The search for trace elements in presolar silicate grains. 85th Annual Meeting of The Meteoritical Society 85

[97] Trappitsch R, et.al. Simultaneous iron and nickel isotopic analyses of presolar silicon carbide grains. Geochimica Et Cosmochimica Acta. 2018;**221**:87–108

[98] Takigawa A, Stroud RM, Nittler LR, Alexander OD, M. C, Miyake A. High-temperature dust condensation around an AGB star: Evidence from a highly pristine presolar corundum. The Astrophysical Journal. 2018;**862**. DOI: 10.3847/2041-8213/aad1f5

[99] Liu N, Lugaro M, Leitner J, Meyer BS, Schönbächler M. Presolar grains as probes of supernova nucleosynthesis. Space Science Reviews. 2024;220. DOI: 10.1007/s11214-024-01122-w

[100] Meech KJ, et.al. A brief visit from a red and extremely elongated interstellar asteroid. Nature. 2017;**552**:378–381

[101] Jewitt D, Hui M-T, Kim Y, Mutchler M, Weaver H, Agarwal J. The nucleus of interstellar comet 2I/Borisov. The Astrophysical Journal. 2020;888. DOI: 10.3847/2041-8213/ab621b

[102] Martinez-Palomera J. And 153 colleagues 2025. Comet 3I/ATLAS. Minor Planet Electronic Circulars. 2025; R63. DOI: 10.48377/MPEC/2025-R63

[103] Rafikov RR. 1I/2017 'Oumuamua-like interstellar asteroids as possible messengers from dead stars. The Astrophysical Journal. 2018;861. DOI: 10.3847/1538-4357/aac5ef

[104] Seligman D, Laughlin G. Evidence that 1I/2017 U1 ('Oumuamua) was composed of molecular hydrogen ice. The Astrophysical Journal. 2020;**896**. DOI: 10.3847/2041-8213/ab963f

[105] Moro-Martn A. Could 1I/'Oumuamua be an icy fractal aggregate? The Astrophysical Journal. 2019;**872**. DOI: 10.3847/2041-8213/ab05df

[106] Deam S, et.al. 2025. A portrait throughout perihelion of the NH2-rich interstellar comet 2I/Borisov. EPSC-DPS Joint Meeting 2025 (EPSC–DPS2025). DOI: 10.5194/epsc-dps2025-1348

[107] Bodewits D and 8 colleagues 2020. Composition and evolution of interstellar comet 2I/Borisov. AAS/Division for Planetary Sciences Meeting Abstracts #52

[108] Hallatt T, Wiegert P. The dynamics of interstellar asteroids and comets within the galaxy: An assessment of local candidate source regions for 1I/'Oumuamua and 2I/Borisov. The Astronomical Journal 159. 2020;**159**:147. DOI: 10.3847/1538-3881/ab7336

[109] Ye Q-Z. Comet 3I/ATLAS. Minor Planet Electronic Circulars. et.al. 2025; N35. DOI: 10.48377/MPEC/2025-N35

[110] Hoogendam WB, et.al. Spatial profiles of 3I/ATLAS CN and Ni

outgassing from Keck/KCWI integral field spectroscopy. arXiv E-prints. 2025

[111] Opitom C, et.al. Snapshot of a new interstellar comet: 3I/ATLAS has a red and featureless spectrum. Monthly Notices of the Royal Astronomical Society. 2025;**544**:L31–L36

[112] Yalgın Ş, Aksaker N, Erzincan N, Akyuz A. Optical spectroscopy and photoionization modeling of four selected planetary nebulae. New Astronomy. 2025;**119**:102413. DOI: 10.1016/j.newast.2025.102413

[113] Aksaker N, Erzincan N, Akyuz A, Temiz U. Physical properties, abundances and photoionization modeling of 12 Galactic PNe. Acta Astronomica. 2024;**74**:201–229. DOI: 10.32023/0001-5237/74.3.2

[114] Nanni A, Cristallo S, Donevski D, Michalowski MJ, Romano M, Sawant P. Dust survival in harsh environments. Investigating the relevance of photo-evaporation process. Astronomy and Astrophysics. 2024;684. DOI: 10.1051/0004-6361/202348024

[115] Hrivnak BJ, Lu W, Bakke WC, Grimm PJ. Variability in protoplanetary nebulae. IX. Evidence for evolution in a decade. The Astrophysical Journal 939. 2022. DOI: 10.3847/1538-4357/ac938a

[116] McFarland J, Speck AK. Dust driven multiphase hydrodynamics in planetary nebulae. NSF Award. 2018;**18**:12946

[117] Delgado-Inglada G, Morisset C, Stasinska G. Ionization correction factors for planetary nebulae - I. Using optical spectra. Monthly Notices of the Royal Astronomical Society. 2014;**440**:536–554. DOI: 10.1093/mnras/stu341

[118] Garcia-Hernández DA, Ventura P, Delgado-Inglada G, Dell'Agli F, Di Criscienzo M, Yagüe A. Galactic planetary nebulae with precise nebular abundances as a tool to understand the evolution of asymptotic giant branch stars. Monthly Notices of the Royal Astronomical Society. 2016;**461**:542–551. DOI: 10.1093/mnras/stw1349

[119] Stanghellini L, et.al. The nature of dust in compact Galactic planetary nebulae from Spitzer spectra. The Astrophysical Journal. 2012;**753**. DOI: 10.1088/0004-637X/753/2/172

[120] Umana G, Trigilio C, Cerrigone L, Buemi CS, Leto P. ATCA observations of the very young planetary nebula SAO 244567. Monthly Notices of the Royal Astronomical Society. 2008;**386**:1404–1410. DOI: 10.1111/j.1365-2966.2008.13044.x

[121] Khushbu K, Muthumariappan C. Photo-ionization structures of planetary nebula IC 2003 with [WR] central star. Advances in Space Research. 2024;**74**:1366–1377. DOI: 10.1016/j.asr.2024.04.058

[122] Galliano F, Galametz M, Jones AP. The interstellar dust properties of nearby galaxies. Annual Review of Astronomy and Astrophysics. 2018;**56**:673–713. DOI: 10.1146/annurev-astro-081817-051900

Chapter 5

Mean Free Path Values of Wave-Particle Interactions in the Venus Ionosheath and in the Venus Upper Ionosphere

Hector Pérez-de-Tejada

Abstract

A calculation is presented to estimate the mean free path values of wave–particle interactions produced through the transport of solar wind momentum to the Venus upper ionosphere. While particle–particle (Coulomb) collisions are expected to be operative throughout most of the low altitude Venus ionosphere, other processes should be active in its upper layers, where the solar wind momentum is delivered to the local plasma through wave–particle interactions. A procedure derived from dissipation phenomena in fluid dynamics (viscosity and thermal conduction) is applied to the solar wind by using its thermal speed and the viscosity coefficient. Values for the mean free path are derived for the Venus ionosheath and also for the plasma in the Venus upper ionosphere. As a whole, the mean free path from wave–particle interactions decreases from high values in the Venus ionosheath to lower values in the upper ionosphere, where it becomes comparable to those produced by Coulombian collisions. Further calculations lead to values of the kinematic viscosity coefficient that are in good agreement with those that have been derived in studies of the momentum equation of the solar wind.

Keywords: solar wind, wave–particle interactions, Venus ionosphere, trans-terminator flow, Mean free path values

1. Introduction

Studies of the motion of the Venus dayside ionosphere to the nightside (trans-terminator flow) lead to consideration of distinctive phenomena that could account for its kinetic momentum flux. Emphasis was first placed on the pressure gradient forces that arise from the different plasma densities present between the dayside and the nightside hemispheres [1]. In the presence of a pressure gradient between both regions, there should be kinetic motion between them. An alternative view was later suggested in terms of the solar wind momentum that is transferred to the flanks of the Venus upper ionosphere [2]. In that case, there is an approximate balance between the altitude-integrated momentum flux of the trans-terminator flow across the Venus upper ionosphere and the deficiency of momentum flux that the solar wind experiences as it streams by its flank regions [2, 3]. From that momentum flux balance, it is possible to estimate the speed of the ionospheric trans-terminator flow, which turns out to be close to the measured values reported from the Pioneer Venus observations [4].

Different from plasma instabilities that generate wave oscillations in a plasma environment [5, 6] it has also been proposed to consider, instead, dissipative phenomena (viscosity and thermal conductivity) that generate wave–particle interactions as the driving source in the behavior of collision less plasmas. This interpretation was first addressed in a study of fluid dynamic streams in terms of particle–particle interactions [7], that can also be applied to environments with alternative interaction processes, as in the solar wind, where there is a different transport of statistical properties. With this purpose, it is possible to derive relations between the thermal speed in a medium and the mean free path that characterize their statistical transport through representative values of the kinematic viscosity and the thermal conductivity coefficients [8].

2. Equations related to dissipative phenomena

Values of the mean free path derived from wave–particle interactions in the solar wind can be obtained from the fluid dynamic momentum and energy equations discussed in the study by Liepmann and Roshko [7] as:

$$\nu = V_T \lambda_H \text{ for viscous dissipation} \tag{1}$$

$$V_T \lambda_H = \alpha\, k/\rho c_p \text{ for thermal dissipation} \tag{2}$$

where V_T is the thermal speed, ν the kinematic viscosity coefficient, and $k/\rho c_p$ the corresponding transport parameter for heat diffusion. Both equations are related to the mean free path λ_H, derived by considering that the media involved is in a mixing process (which can be applied to the solar wind and the Venus upper ionosphere). The flow only experiences small deviations away from equilibrium, and thus we will assume that there are linear relations between stress and the rate of strain for viscous dissipation, and also between the heat flow and the temperature gradient in the case of thermal dissipation. The purpose of the approach is to obtain the gas dynamic expressions for the shear viscosity μ and for the heat conductivity k when the flow is subject to wave–particle interactions.

Measurements conducted with spacecraft in the Venus wake lead to variables that are involved in the above equations. In particular, the thermal speed V_T of the solar wind is reported from the Mariner 5 spacecraft plasma data obtained during its flyby crossing through the Venus wake, and this is reproduced in the upper panel of **Figure 1** [9]. The measured values show that, in general terms, V_T varies between 60 km/s in the solar wind and -100 km/s within the ionosheath. Separately, smaller speed values for the planetary O^+ ions in the trans-terminator flow are shown in **Figure 2** [10], where they only reach 3–4 km/s. The speed of the solar wind protons is related in Eq. (1) to values of the kinematic viscosity coefficient ν that have been derived from the momentum equation of the solar wind as it streams along the Venus ionosheath. In fact, since the freestream magnetic field intensity is $B \sim 10$ nT in the first panel of **Figure 1** [9], the $J \times B$ force can be neglected, as it becomes much

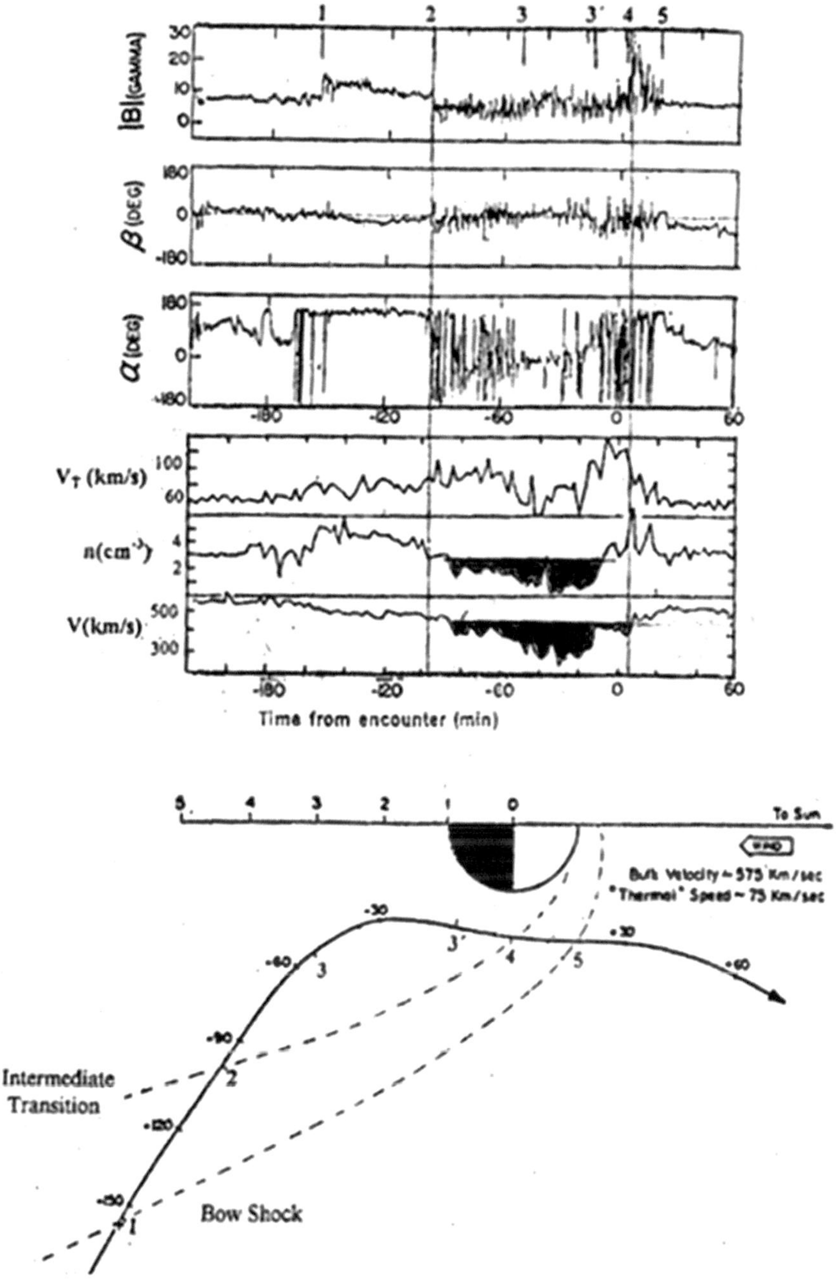

Figure 1.
Lower panel: Trajectory of the Mariner 5 spacecraft projected in cylindrical coordinates during its flyby past Venus. Labels 1 to 5 along the trajectory mark important events in the plasma properties (a bow shock is identified at features 1 and 5, and the intermediate plasma transition occurs at features 2 and 4). Upper panel: Magnetic field intensity and its latitudinal and azimuthal orientation, together with the plasma properties (thermal speed, density, and bulk speed) measured around Venus [9].

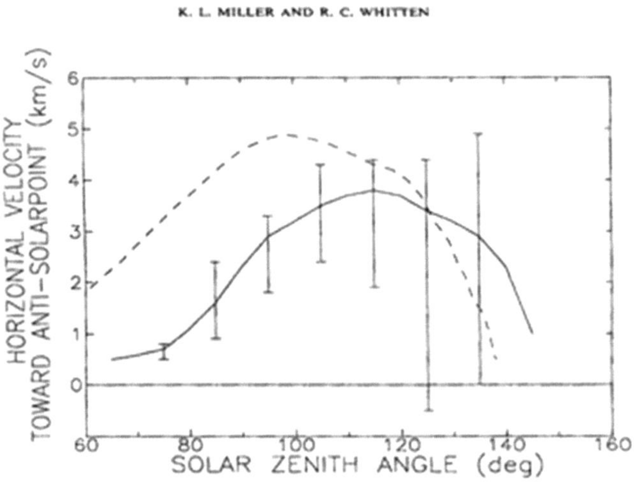

Figure 2.
Approximate range of speed values of the trans-terminator flow of planetary O^+ ions measured with the Pioneer-Venus Orbiter (PVO) spacecraft in the Venus upper ionosphere as a function of the solar zenith angle (the peak speed values reach 3–4 km/s) [10].

smaller than the viscous force applied across the velocity shear that is represented in **Figure 3** [11] from the outer boundary of the wake. From such considerations, it is possible to derive values for the kinematic viscosity coefficient of the solar wind as $\nu \sim 3\ 10^5\ km^2/s$ [4, 8].

Changes in the λ_H value can also be expected from differences in the ion population, with large values in the speed of the solar wind protons that are dominant in the ionosheath, and then low-speed (3–4 km/s), high-density planetary ions O^+ in the upper ionosphere, as depicted in **Figure 2** [10]. In addition, it should be noted that the cross-section of the plasma flow in the ionosheath is larger than that in the upper ionosphere. This property can be appreciated by comparing the ≤ 1,000 km altitude of the ionopause by the terminator that is schematically indicated in **Figure 4** [1], with position 2 in **Figure 3** [11] (labeled IT) for the outer extent of the boundary layer that reaches large ≥ 1,000 km distances from Venus in the data of the Venera 10 spacecraft trajectory across the Venus wake (reproduced at the top of that figure). That layer yields a much larger distance away from the central region of the wake [11] and that is comparable to the curve labeled intermediate transition in the bottom panel of **Figure 1** [9]. As a result, the cross-section of the region where both particle populations move is different and hence the plasma conditions will not be the same in the upper ionosphere as that in the ionosheath. The implication here is that as noted above the mean free path for wave–particle interactions should be smaller in the upper ionosphere than those in the ionosheath. A further constraint develops at even lower altitudes within the ionosphere with smaller values of the mean free path for particle–particle interactions that become dominant with increasing density of the O^+ ion population.

With such numbers, the above Eq. (1) for viscous dissipation leads to $\lambda_H \geq 10^3$ km values as an approximate estimate for the mean free path suitable for wave–particle interactions of the solar wind in the Venus ionosheath. Varying λ_H values can also be obtained if the thermal speed V_T is larger in the ionosheath. Depicted

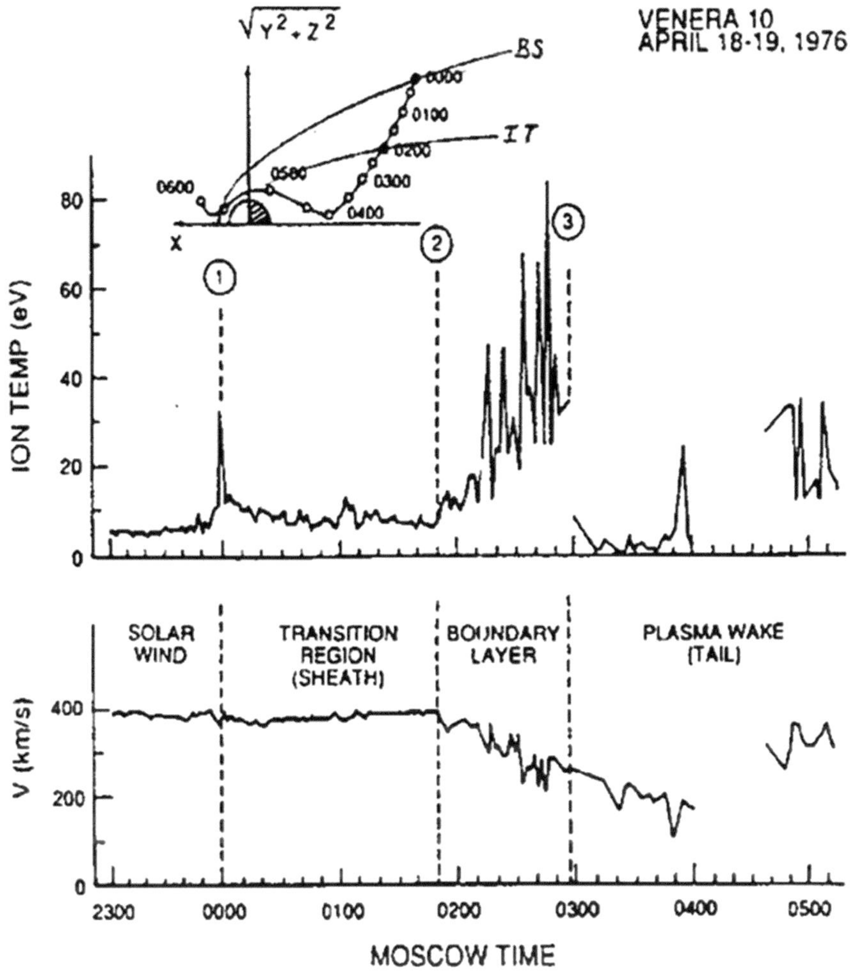

Figure 3.
Ion speed and temperature measured along the orbit of Venera 10 on April 19, 1976. The Venera orbit in cylindrical coordinates is shown at the top. The temperature burst at position 1 was recorded during a flank crossing of a bow shock. A boundary layer is indicated by the increase in temperature and the decrease in speed, and is initiated by the intermediate transition at position labeled 2. A later discontinuity in the boundary layer temperature profile corresponds to the boundary of the Magneto-tail [11, 12].

values of that parameter are illustrated in **Figure 5** [8], showing how λ_H can be larger in the solar wind where V_T is smaller.

3. Calculation of viscous forces

Since the solar wind can account for the trans-terminator flow in the Venus upper ionosphere through momentum transfer, it is necessary to estimate the manner in which that process can be conducted. In particular, the nearly equal values between the deficiency of the momentum flux of the solar wind that streams around the Venus ionosphere and the momentum flux of the trans-terminator flow, as reported

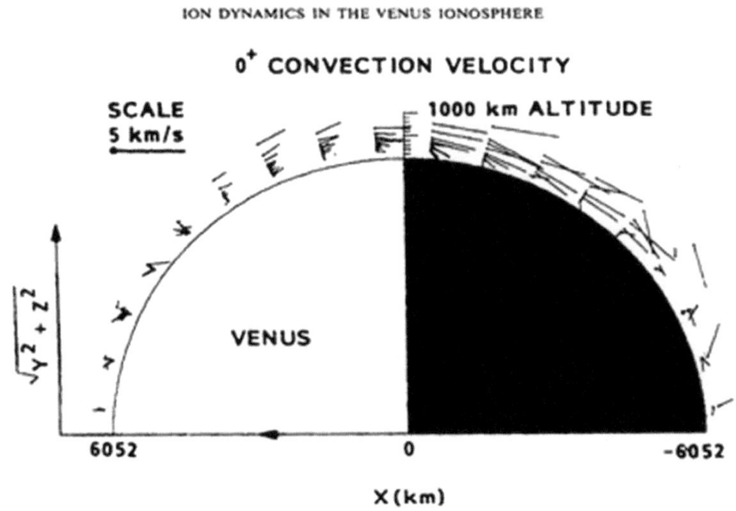

Figure 4.
Vector velocity speeds of the trans-terminator flow in the Venus upper ionosphere, bounded by the ionopause near 1,000 km altitude and measured with instruments onboard the Pioneer Venus Orbiter spacecraft [1].

from the Pioneer Venus Orbiter (PVO) measurements [1, 10], make it possible to derive the latter values through momentum transfer processes. Following this procedure, we can explore how the momentum of the trans-terminator flow is distributed to the lower ionosphere [2]. In this sense, it is necessary to resolve how the kinetic motion of the trans-terminator flow is shared with the low-altitude ionosphere as the ultimate receiver of the solar wind momentum.

An account of this procedure can be elaborated by considering that the kinetic energy of the trans-terminator flow can now be employed as being distributed to the lower ionosphere; in other words, there should be a balance between that energy source given in terms of the inertial forces of the trans-terminator flow and the amount received at lower altitudes through viscous forces. Thus, we will address the momentum equation of the trans-terminator flow:

$$\rho U \partial U / \partial x = \mu \partial^2 U / \partial y^2$$

Given in terms of viscous forces through the viscosity coefficient μ along the *y*-direction transverse to that of the flow motion in the *x*-direction. From this equation, we can write: $U \Delta U / L = \nu \Delta U / \delta^2$ where $\nu = \mu/\rho$ is the kinematic viscosity coefficient of the flow, and δ is the width of a viscous boundary layer of length L (6×10^3 km ~ 1 Venus radius).

Values of these variables are available from the plasma data obtained during the crossing of the Venus wake by the Mariner 5 spacecraft, as reproduced in **Figure 1**. From the upper panel of that figure, where the thermal and the bulk speed values are respectively labeled as V_T and V, we can take V =500 km/s for the freestream solar wind speed, and from the lower panel, take δ = 2,000 km for the distance between the ionopause and the features labeled 3 and 4 in the Mariner 5 trajectory, which identify the outer extent of the velocity boundary layer by the terminator plane. With these numbers, we obtain $3 \times 10^5 \sim \nu$ km^2/s^{52}

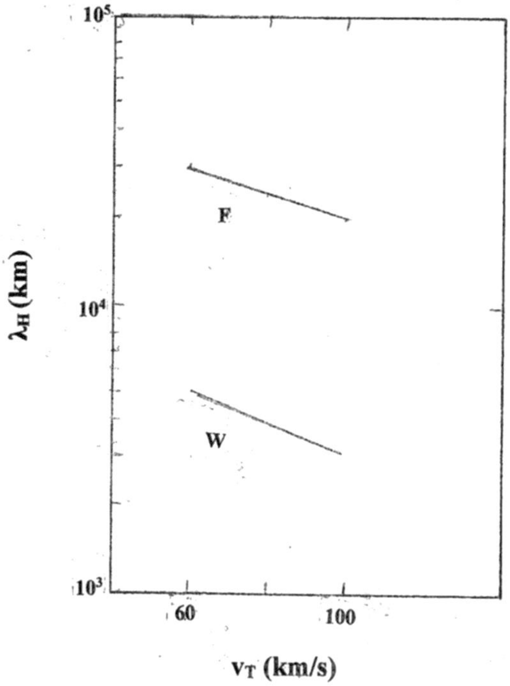

Figure 5.
*Mean free path values, λ_h, of the solar wind obtained through wave–particle interactions and magnetic field fluctuations, using the solar wind thermal speed, V_T, and the kinematic viscosity coefficient, ν, measured during the Mariner 5 trajectory shown in **Figure 1**. The connecting line labeled "W" refers to values in the Venus ionosheath and in the solar wind that are implied by wave–particle interactions. The connecting line labeled "F" is implied by the magnetic field fluctuations. The implication here is that, for fixed ν values, λ_h becomes smaller with increasing V_T values, as implied in Eq. (1) [8].*

ν ~ from the above relation, which is comparable to that derived by calculating the Reynolds number suitable to the solar wind that streams by the Venus ionosheath, as implied from measurements conducted with the Mariner 5 spacecraft shown in **Figure 1** [4, 13].

The similar value of the kinematic viscosity coefficient of the solar wind, as it interacts with the Venus ionosphere obtained here and that noted above in Section 2, and that is derived from the study of its momentum equation [14], stresses its validity within that value range.

4. Discussion

From the mean free path value of wave–particle interactions derived for the plasma within the Venus upper ionosphere and, at the same time, its larger values derived for the ionosheath flow in the interaction region with the solar wind, it is possible that their values may be related to the different scale size of the region where they apply. In fact, since the width of the Venus ionosheath ($\lambda_H \geq 10^3$ km), implied from the shape of the boundary layer in **Figure 3** [11, 12] may reach a few times the value within the ionosphere ($\lambda_H < 10^3$ km), as implied by the width of the trans-terminator flow in **Figure 4** [1], it is possible that the larger plasma densities of the plasma conditions within the upper ionosphere may lead to values of the mean free path in the upper ionosphere that are smaller than those in the ionosheath and thus lead to smaller values of the mean free path in the upper ionosphere.

Differences in the value of the mean free path λ_H with respect to the thermal speed of the speeding flow are suggested in the profiles presented in **Figure 5** [8] ($V_T \sim 60$ km/s and 100 km/s) and correspond, respectively, to values in the solar wind and in the ionosheath flow. On the contrary, the λ_H values in the ionosheath may be up to one order of magnitude larger than the λ_H values in the upper ionosphere. The latter is reminiscent of the smaller values expected from Coulombian conditions at lower altitudes in the Venus ionosphere (see discussion after Eq. (6) in the study [14]). As a whole, it can be argued that wave–particle interactions that are operative in the ionosheath may extend their domain to the Venus upper ionosphere, where they are gradually replaced by Coulombian collisions at lower altitudes [14].

A similar difference may also be relevant for plasma conditions present in other plasma encounters in astrophysical problems. In particular, if the width of an ionosheath flow external to a planetary ionosphere is much larger than the width of the ionosphere, the difference in their mean free path values should be more evident. At the same time, interaction problems between oncoming directional plasma flows and rotating galactic structures could also lead to large differences in their corresponding mean free path values.

5. Conclusions

From the equations that relate dissipation phenomena to the mean free path of wave–particle interactions in the solar wind, it is inferred that large λ_H values can be encountered in the ionosheath flow and that they should gradually decrease toward the ionosphere. The momentum equation of the trans-terminator flow in the Venus ionosphere maintains a balance between its inertial forces and the momentum contents that are delivered to the lower-altitude ionosphere.

Author details

Hector Pérez-de-Tejada
Institute of Geophysics, Unam, Mexico

*Address all correspondence to: hectorperezdetejada@gmail.com

References

[1] Knudsen WC, Spenner K, Miller KL, Novak V. Transport of Ionospheric O^+ ions across the Venus terminator. Journal of Geophysical Research. 1980;**85**:7803. Antisolar acceleration of ionospheric plasma across the Trans-terminator, Geophys. Res. Lett.8. 241,**1981**

[2] Pérez-de-Tejada H. Fluid dynamic constraints of the Venus Ionospheric flow. Journal of Geophysical Research. 1986;**91**:6765–6770

[3] Lundin R, et al Ion flow and momentum transfer in the Venus plasma environment. Icarus. 2011;**215**:751–758

[4] Pérez-de-Tejada H. Vortex Structures in Planetary Plasma Wakes. London, UK: Cambridge Scholars Pub: Cambridge, UK: IntechOpen; 2023

[5] Szegö K, Shapiro VS, Shevchenko VI, Sagdeev RZ, Kasprzak WT, Nagy AF. Physical processes in the plasma mantle of Venus. Geophysical Research Letters. 1991;**18**:2305

[6] Delva M, Zhang TL, Volwerk M, Vörös Z, Pope SA. Proton cyclotron waves in the solar wind at Venus. Journal of Geophysical Research. 2008;**113**:E00B06

[7] Liepmann J, Roshko A. Elements of Gasdynamics. Hoboken, NJ, USA: John Wiley; 1957. p. 372

[8] Pérez-de-Tejada H.Wave particle interactions in Astrophysical. Plasmas Galaxies. 2024;**12**(3):28

[9] Bridge HS, Lazarus AJ, Snyder CW, Smith EJ, Davis L, Coleman PJ, Jones DE, Mariner V. Plasma and Magnetic fields observed near Venus. Science. 1967;**158**:1669–1673

[10] Miller K, Whitten R. Ion dynamics in the Venus ionosphere. Space Science Reviews. 1991;**55**:165

[11] Romanov SA, Smirnov V, Vaisberg O. Interaction of the solar wind with Venus. Cosmic Research. 1979;**16**:603

[12] Phillips JL, McComas DJ. The magnetosheath and magnetotail of Venus. Space Science Reviews. 1991;**55** (1):80

[13] Pérez-de-Tejada H. Viscous forces in velocity boundary layers around planetary Ionospheres. Astrophysical Journal. 1999;**525**:L65–L68

[14] Pérez-de-Tejada H, Reyes-Ruiz M, Manterola HD. Viscous flow properties in the transport of solar wind momentum to the Venus upper ionosphere. ICARUS. 2010;**206**:182–186